Practical Structural Analysis for Architectural Engineering

Practical Structural Analysis for Architectural Engineering

August E. Komendant, D.E., P.E.

Consulting Engineer
Dist. Visiting Professor

PRENTICE-HALL, INC.
Englewood Cliffs, New Jersey 07632

Library of Congress Cataloging-in-Publication Data

Komendant, August E. (date)
  Practical structural analysis for architectural engineering.

  Includes index.
    1. Concrete construction.   2. Structures, Theory of.
I. Title.
TA681.5.K66 1987   624.1′834   86-5029
ISBN 0-13-693961-9

Editorial/production supervision and
  interior design: Lynda Griffiths
Cover design: Wanda Lubelska
Manufacturing buyer: Rhett Conklin

© 1987 by August E. Komendant

All rights reserved. No part of this book may be
reproduced, in any form or by any means,
without permission in writing from the publisher.

Printed in the United States of America

10 9 8 7 6 5 4 3 2 1

ISBN 0-13-693961-9 025

PRENTICE-HALL INTERNATIONAL (UK) LIMITED, *London*
PRENTICE-HALL OF AUSTRALIA PTY. LIMITED, *Sydney*
PRENTICE-HALL CANADA INC., *Toronto*
PRENTICE-HALL HISPANOAMERICANA, S.A., *Mexico*
PRENTICE-HALL OF INDIA PRIVATE LIMITED, *New Delhi*
PRENTICE-HALL OF JAPAN, INC., *Tokyo*
PRENTICE-HALL OF SOUTHEAST ASIA PTE. LTD., *Singapore*
EDITORA PRENTICE-HALL DO BRASIL, LTDA., *Rio de Janeiro*

Contents

PREFACE   ix

1 FUNDAMENTALS IN STRUCTURAL ANALYSIS   1

    General   1

    Assumptions   2

    Basics for Structural Analysis   3

    Data and Rules   4

    Simple Beams   9
    Reactions, Shear, Moments, Deformations, 9

2 STATICALLY INDETERMINATE BEAMS   14

    Method of Analysis   14
    Zero-Point Method, 15
    Loading Factor Method, 19
    Span-by-Span Loading, 24

    Support Moments   26
    Uniform Loading, 27
    Concentrated Loads $P = 1.0$, 33

Critical Live Load Locations   34

Moments and Reactions in Short Span Slabs   38

3 ONE- AND MULTISTORY FRAMES 40

General   40
Zero-Point Method, 40
Loading Factor Method, 46

Multistory Frames   47
Uniform Vertical Loading, 47
Column Moments, 49
Concentrated Loads, 54
Horizontal Loading, 55

Lateral Deflection   65

One-Story Multibay Frames   67

Temperature Change and Shrinkage   68

Elastically Controlled Joints   71

Simple Frames   72
General, 72

4 TRUSSES 83

General   83

Simple Trusses   83

Vierendeel Girders   86
Simplified Analysis, 86

5 TWO-WAY SLABS AND GRIDS 97

Method of Analysis   97
One-Span Slabs, 99
Continuous Slabs or Grids, 103
Slab Moments, 106
Framing, 109
Beams in y-Direction, 111

Flat Slabs   112

Two-Skin and Waffle Slabs   114

Contents   vii

6 DIMENSIONING AND ULTIMATE CARRYING CAPACITY   116

General   116
Allowable Stresses, 118
Recommended Slenderness Limits, 118

Simple Bending   119

Combined Bending and Axial Normal Force   123
Large Eccentricity, 124
Small Eccentricity, 127

Ultimate Carrying Capacity and Factor of Safety   128
General, 128

7 PRESTRESSED CONCRETE   132

Basic Principles   132

Application of Prestressing   135
Suspension Action, 135

Statically Determinate Prestressed Members   137

Statically Indeterminate Prestressed Members   139
Two Spans (EI = constant), 139
Three Spans, 143

Factor of Safety   145

Loss in Prestress   147
Initial Losses, 147
Loss in Prestress Due to Shrinkage and Creep, 148

Deflection   158

8 ARCH ACTION   160

General   160

Three-Hinged Arches   164
Deflections of Crown, 165
Shape of Arch, 166

Two-Hinged Arches   167

Fixed Arches   172

9 SPACE STRUCTURES   176

General   176
Loading, 176

Domes   177
Membrane Forces, 177
Marginal Member, 181
Deformations, 181

Folded Plates   184
Simplified Theory, 184

Curvilinear Cylindrical Shells (Barrels)   187
Membrane Theory, 187
Membrane Forces, 189
Edge Member, 192
Beam Theory, 193

Polygonal Domes   199
Membrane Theory, 199
Marginal Member, 201

Hyperbolic Paraboloids   205
Membrane Theory, 205
Membrane Forces, 207
Edge Members, 207

10 STRUCTURES SUBJECTED TO DYNAMIC FORCES   210

General   210
Code Method, 211
Energy Method, 212

Wind Loading   215

Column Stresses   216

11 FOUNDATIONS AND RETAINING WALLS   221

General   221
Strip and Mat Footings on Semielastic Soil, 222

Retaining Walls   226
Free Standing Walls, 227
Laterally Supported Walls, 231
Basement Walls Partly Below Grade, 236

APPENDIX   241

INDEX   253

Preface

As a professor and consulting engineer-designer working with students, practicing architects, and engineers, I find a considerable lack of knowledge and understanding about structural systems, related materials, and their behavior under conceivable loading.

Many colleagues and students here and abroad strongly requested and tried to convince me to use my experience in practice and in teaching to write a book about the design of concrete structures that would serve both professions—architecture and engineering. This idea lingered in my mind for a long time. I studied a series of books in this field, most of which required a considerable knowledge of mathematics, which the architects, most engineers, and students do not have or have forgotten. Most texts available have overlooked application of the theories to practical design and do not present clearly enough the structural principles and deformations under loading conditions. Proper understanding of structures and basic structural principles is the prerequisite for studying engineering. In addition, aesthetics is commonly overlooked. Aesthetics, novelty, and efficiency are the vital ingredients that make a design significant.

As a result, in practice many structural designs in concrete are based on rough approximation without knowing or considering the limitations of their application. Therefore, bad concrete designs exist and failures occur.

Any designer—whether an architect or engineer in practice—must be up to date and know what is available to select a proper structural system for given conditions. Without such knowledge, the results are never satisfactory and reveal struggle between hope and despair in every aspect.

Taking all of this into consideration, I decided to write a book on structural analysis for concrete structures to fill the gap between rather complex advanced structural theories and their application in practice.

Such a book must be as simple as possible, requiring no mathematics beyond the elementary level. Regardless of simplifications, the structural analysis presented must describe the stress condition and deformations of structures subjected to common loadings quantitatively and qualitatively within acceptable limits.

Easily understandable methods for structural analysis are available based on rigorous theories; but these are simplified for practical application. Such methods are needed by practicing engineers and even architects, as well as engineering and architectural students. Easily understandable, applicable, and appealing methods raise interest in students, and they will not forget what they have learned.

As can be seen from the contents of the book, all methods used have been proven in practice. The required data for application are given. Applying such methods, the redundants—support moments—of statically indetermined structural systems, such as continuous beams and frames, are obtained very quickly with sufficient accuracy by using only a pocket-size calculator.

In cases where such simplified methods are not available, such as for simple frames, arches, shells, and so on, close formulas point out their limits in application.

Proper attention has been paid to structural problems, such as temperature change, shrinkage, ultimate carrying capacity, and factor of safety of structures subjected to severe temporary loadings, which most existing texts overlook or cover inadequately.

All methods used are illustrated for clarification by numerical examples from practice. The appendix contains tables for soil physical characteristics and elementary transcendental functions required to determine the losses in prestressing and foundation designs on semielastic soils. Also, tables for moment coefficients for continuous beams and two-way slabs, standard reinforcing bars, welded wire fabrics, and prestressing strands are given with required structural data.

Since the metric system used is still relatively new to Americans, the conversion table to psi- and SI-systems and vice versa is added.

In conclusion, the highest credit and my deepest gratitude belong to my professors Kurt Beyer and Benno Löser, who have taught me the theories and methods I have successfully applied for about forty-five years and which pervade this book. I am sincerely thankful to professor David E. Guise, who read the manuscript and gave me valuable advice, and to my daughter, architect Merike Phillips, for the numerous illustrations. Also, I appreciate the excellent work of Prentice-Hall's editorial and production staff.

August E. Komendant
Upper Montclair, New Jersey

Practical Structural Analysis for Architectural Engineering

1

Fundamentals in Structural Analysis

GENERAL

Structural analyses are based on assumptions. Without assumptions, the analyses for reinforced concrete would be too complex or even practically not manageable. Since concrete is not a homogeneous material, its volume changes over the course of time. Since it is manufactured in field, its strength is not uniform and depends in large degree on curing and climatic conditions. Its tensile strength is only a fraction of its compressive strength. The compressive strength and related modulus of elasticity (E_c) is a function of time. Thus, the ratio of the modulus of elasticity steel to concrete $E_s/E_c = n$ may vary from 6 to 15 under normal conditions.

The design commonly is carried out for finished structural systems, but in construction the structural system changes from floor to floor. In accordance with experience, the most critical condition for the structure occurs during construction, because the structural system is usually less efficient and loading may be heavier than the design load.

For simplification, beams or girders supported by columns are assumed and designed to be continuous beams but they actually are acting as frames. The shrinkage and temperature stresses are highest during construction. The columns resist the related changes of beam length and are subjected to rather high moments and related flexural stresses. The efficiency of flexural stresses in comparison to central compressive stresses is only about 15 percent, which leads to heavier columns and inefficient use of concrete.

It must be realized that the exact theories for reinforced concrete structures are extremely complex; therefore, very often in practice a design is based on unrealistic or even wrong methods and assumptions for simplicity or convenience. Also, the field conditions are very often not as assumed. In order to carry out a design with given time and expenditures, simplification and assumptions are justified, but they have to be realistic. If this is not the case, it may lead to failures and even fatalities.

In the following chapters, we will discuss these points and show how the difficulties can be overcome in the simplest way and how the gap between theory and practice can be closed to obtain results that are acceptable quantitatively and qualitatively.

ASSUMPTIONS

The structural analysis of reinforced concrete is based on the following generally approved assumptions.

1. Concrete is elastic and obeys Hooke's law, which means that the flexural stresses (σ) are proportional to the strain (ε) and inversely proportional to the modulus of elasticity of concrete (E)

$$\sigma = E\varepsilon$$
$$\varepsilon = \frac{\sigma}{E} \qquad (1\text{-}1)$$

and the stress distribution is linear over the cross-sectional area (Eq. 1-1).

2. The flexural stresses (σ_c) in a homogeneous cross section (prestressed concrete) and in an unhomogeneous cross section (reinforced concrete) are proportional to the moment (M) and distance (z) from the center of gravity (C.G.) and from the neutral plane (N.P.), respectively, and are inversely proportional to the moment of inertia (I). This relation is expressed by Eq. 1-2.

$$\sigma_c = \frac{M}{I} z \qquad (1\text{-}2)$$

3. The tensile strength of concrete is rather small in comparison to compressive strength. It is disregarded in stress analysis and is substituted by reinforcing steel as an integral part of the section. Thus, the steel stress is (Eq. 1-3)

$$\sigma_s = \frac{E_s}{E_c} \sigma_c = n\sigma_c \qquad (1\text{-}3)$$

4. Due to the linear elastic behavior of stresses, the external forces—reaction (R), shear (V), and moments (M) caused by a combination of loadings—can be treated separately and then can be superimposed from each loading type.

BASICS FOR STRUCTURAL ANALYSIS

Equilibrium of a rigid body at rest requires that the sum of all forces that the body is subjected to in horizontal (X) and vertical (Z) directions must be zero at any point and the moment (M) of these forces about the center of gravity is also zero to avoid rotation of the body (Eq. 1-4). Thus,

$$\Sigma X = 0$$
$$\Sigma Z = 0 \qquad (1\text{-}4)$$
$$\Sigma M = 0$$

These three requirements are called "three equations of equilibrium" and serve as a basis for structural analysis.

To obtain a clear understanding of the basic assumption and equilibrium requirements, the reinforced concrete section illustrated in Fig. 1-1 will be considered. The section is subjected to external loading (w), moments (M), and shear (V), which have to be in equilibrium with the resultants (C) and (T) of flexural stresses (σ), shear (v), and bonding stresses (u). These stresses should not exceed the safety limits or allowable stresses. For example, the bonding between steel and concrete should not be interrupted. If this occurs, steel will not any more be an integral part of the section and the theory of the design is no longer valid. Such failure results in cracking, extensive sagging of the beam, and finally leads to collapse.

Besides these requirements, there are two others: acceptable deflection and no vibrations in the structure under normal conditions.

The final criteria of the quality of a design is that it must have a margin of safety for any extreme conceivable loading that the structure may be subjected to. The factor of safety (F.S.) commonly is 1.8 times the design loading.

Thus, the structural analysis consists of two parts: (1) the determination of the external moments, reaction, and shear for given or estimated loading (w., P.); (2) dimensioning of the critical sections and computation of the stresses and deflection (δ).

The relationships of external and internal forces (stresses) are (Fig. 1-1):

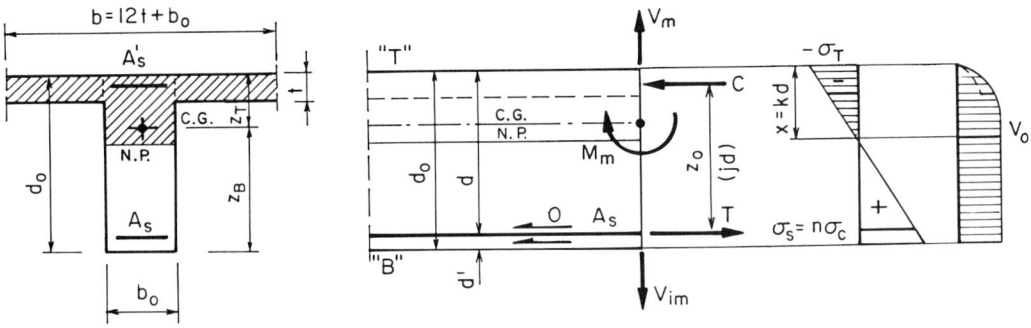

Figure 1-1

$$V_m + V_{im} = 0 \qquad V_m = R - x_m w$$

$$T + C = 0 \qquad T = -C$$

$$-Cz_o = Tz_o = M_i \qquad -M_i = M_m$$

$$T = -C = \frac{M_m}{z_o} \tag{1-5}$$

$$V = \frac{\Delta M}{\Delta x}$$

$$\frac{\Delta T}{\Delta x} = -\frac{\Delta C}{\Delta x} = \frac{\Delta M}{z_o \Delta x} = \frac{V}{z_o} = vb_o = \Sigma ou$$

$$\frac{\Delta V}{\Delta x} = w = \frac{\Delta^2 M}{\Delta x^2}$$

where z_o is the lever arm of the resultants C of compressive stresses (σ_c), T is the tensile stress in steel (σ_s), and (o) is the perimeter of reinforcing in contact with the concrete. The steel stress (σ_s), concrete stress (σ_c), shear stress (v), and bond stress (u) are

$$\sigma_s = \frac{T}{A_s}, \qquad A_s = \frac{T}{\sigma_s}$$

$$\sigma_c = \frac{2T}{k\,d\,b_o}, \qquad T = -C, \qquad k = \frac{n}{n+m} \tag{1-6}$$

$$v = \frac{V}{b_o z_o} \qquad\qquad m = \sigma_s/\sigma_c$$

$$u = \frac{vb_o}{\Sigma o}$$

wherein m is the ratio of allowable steel stress (σ_s) to concrete (σ_c).

DATA AND RULES

For any structural analysis, cross-sectional and material coefficients are commonly used in all carrying actions. They are cross-sectional area (A), static moment (Q), moment of inertia (I) about the center of gravity (C.G.) or neutral plane (N.P.), thermal coefficient of expansion (α_t), modulus of elasticity (E), and modulus of shear (G). For concrete, the cross-sectional coefficients for members are determined as shown in Fig. 1-2.

For convenience, the moment of inertia (I) is computed for the top fiber of the section and then is transferred to the center of gravity or neutral plane by the formula

Data and Rules

in Eqs. 1-7 to 1-12, where z_T is the distance from the top to the center of gravity or neutral plane.

$$I_o = I_T - z_T^2 A$$

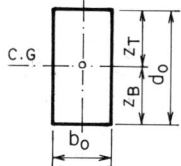

Figure 1-2(a)

$$A = b_o d_o$$

$$z_T = z_B = d_o/2 \tag{1-7}$$

$$I_T = \frac{1}{3} b_o d_o^3 \qquad I_o = I_T - z_T^2 A$$

$$= \frac{1}{12} b_o d_o^3$$

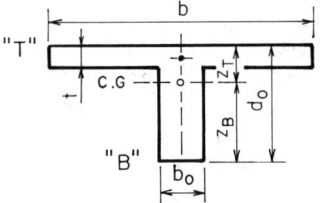

Figure 1-2(b)

$$b = 12t + b_o$$

$$\begin{array}{c|c}
b_o d_o = \text{------} \text{ cm}^2 & \dfrac{b_o d_o^2}{2} = \text{--------} \text{ cm}^3 \\[2ex]
(b - b_o)t = \text{------} \text{ ''} & (b - b_o)\dfrac{t^2}{2} = \text{------} \text{ ''} \\[1ex]
\hline
A = \text{------} \text{ cm}^2 & Q_T = \text{------} \text{ cm}^3
\end{array}$$

$$z_T = \frac{Q_T}{A}, \qquad z_B = d_o - z_T \tag{1-8}$$

$$I_T = \frac{b_o d_o^3}{3} + \frac{(b - b_o)t^3}{3} \text{ cm}^4$$

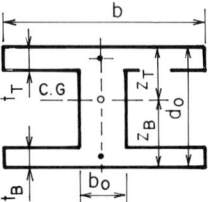

Figure 1-2(c)

$A_1 = b_o d_o = \text{------} \text{ cm}^2 \quad\quad A_1 d_o/2 = \text{--------} \text{cm}^3$

$A_2 = (b - b_o)t_T = \text{-----} \quad " \quad\quad A_2 t_T/2 = \text{--------} \quad "$

$A_3 = (b - b_o)t_B = \text{-----} \quad " \quad\quad A_3(d_o - t_B/2) = \text{------} \quad "$

$\quad\quad\quad A = \text{------} \text{ cm}^2 \quad\quad\quad\quad\quad Q_T = \text{------} \text{ cm}^3$

$$z_T = \frac{Q_T}{A} \quad\quad z_B = d_o - z_T \tag{1-9}$$

$$I_T = A_1 d_o^2/3 + A_2 t_T^2/3 + A_3 t_B^2 + A_3(d_o - t_B/2)^2 \text{ cm}^4$$

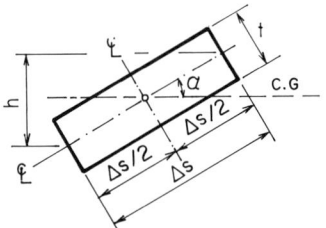

Figure 1-2(d)

$A = \Delta s \, t$

$$I_o = \frac{A(\Delta s^2 \sin^2\alpha + t^2 \cos^2\alpha)}{12} \sim \frac{Ah^2}{12} \tag{1-10}$$

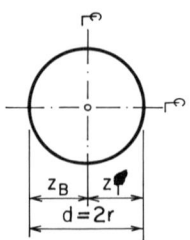

Figure 1-2(e)

Data and Rules

$$z_T = \frac{\Delta s \sin\alpha + t\cos\alpha}{2} = z_B = h/2$$

$$A_o = \frac{\pi d^2}{4} = \pi r^2 \qquad \pi = 3.14$$

$$I_o = \frac{\pi d^4}{64} = \frac{\pi r^4}{4}$$

Cylinder (1-11)

$$\Delta A = A_o - \frac{\pi(d-2t)^2}{4}$$

$$I = I_o = \frac{\pi(d-2t)^4}{64}$$

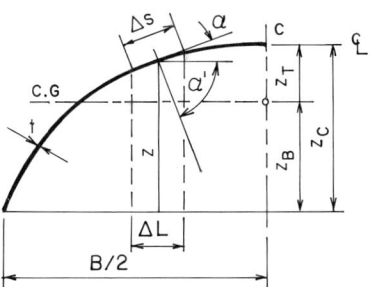

Figure 1-2(f)

$$A = \sum_o^c \Delta s\, t = \sum_o^c \Delta A \qquad \sum A = 2A$$

$$Q_B = \sum_o^c \Delta s\, t\, z = \sum_o^c \Delta A z \qquad \sum Q_B = 2Q_B$$

$$z = 4z_c \xi \xi' \qquad \text{(parabola)}$$

$$\xi = \frac{x}{B} \qquad \xi' = 1 - \xi \tag{1-12}$$

$$z_B = \frac{Q_B}{A} \qquad z_T = z_c - z_B$$

$$I_B = \Sigma(\Delta I_0 + \Delta A\, z^2) \sim \Sigma \Delta A\, z^2$$

$$I_o = I_B - z_B^2 A$$

The temperature expansion α_t for high-strength concrete is approximately $0.8 \cdot 10^{-5}$. The modulus of elasticity by ROS* formula is

$$**E_c = 5.5 \cdot 10^5 \frac{\sigma_{uL}}{150 + \sigma_{uL}} \text{ kg/cm}^2 \qquad (1\text{-}13)$$

where σ_{uL} is the cylinder stress of concrete. The shear modulus $G = E_c/2$. Modulus of elasticity for reinforcing steel is $E_s = 2.1 \cdot 10^6 \text{kg/cm}^2$ — for strands approximately $E_{ss} = 1.9 \cdot 10^6 \text{kg/cm}^2$.

To avoid confusion and to obtain clarity and structural understanding, the following set of rules are applied:

1. The reactions (R), moments (M,X), loading factors ($\overline{L}$), and end rotations (φ) at the left end of a member are noted by an apostrophe (') and the right end by (").
2. At a joint of different members, the angle between them remains unchanged after deformations.

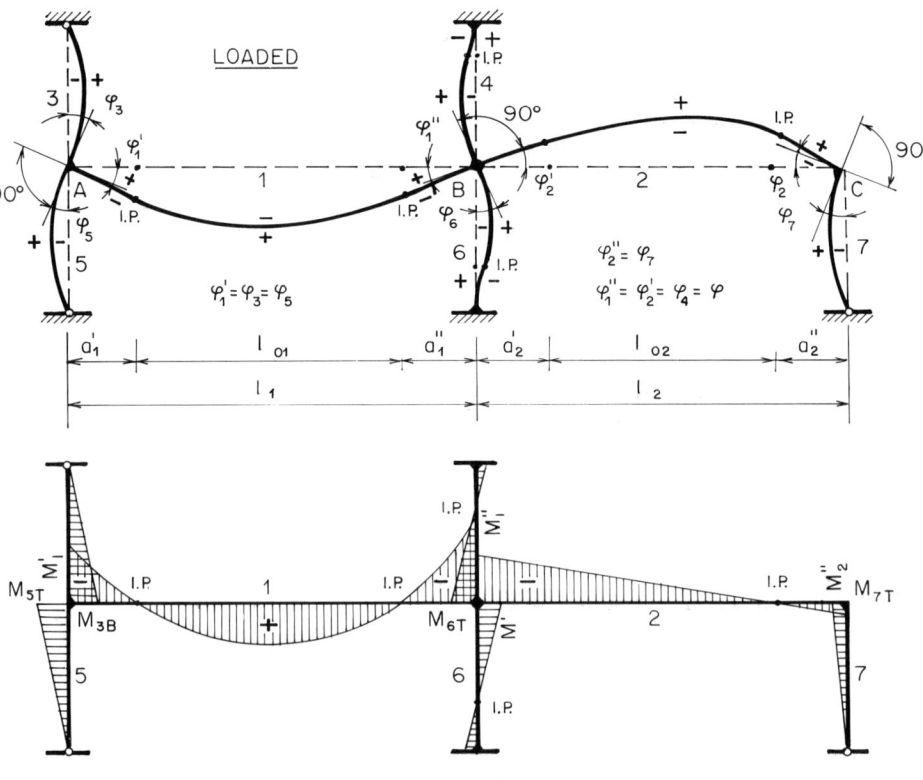

Figure 1-3

*ROS — Tech. University, Zurich, Switzerland.

**The quality of light-weight aggregate varies; therefore, E-value has to be determined by tests for each particular case.

Simple Beams

3. The moments are plotted at the tension side (+) of the member.
4. The top of a vertical member is noted by index $(_T)$ and the bottom by index $(_B)$.
5. Indices: In general, the first index indicates identity, origin, location, or place under consideration; the second index indicates cause, direction, or time of the action.

The rules are illustrated in Fig. 1-3.

SIMPLE BEAMS

Reactions, Shear, Moments, Deformations

The ends of simply supported beams are free to rotate and move. The reactions and moments can be computed from the three equations of equilibrium. In accordance with Otto Mohr,* the end rotation ($EI\varphi,'$ φ'') equals the reactions $\overline{R}'$, $\overline{R}''$ of a beam loaded with the moment (M) and deflection ($EI\delta$) of the moment $\overline{M}$ due to the moment loading (M).

For the most commonly encountered loadings, the reactions (R), shear (V), moments (M), rotations ($EI\varphi$), and deflections ($EI\delta$) will be computed.

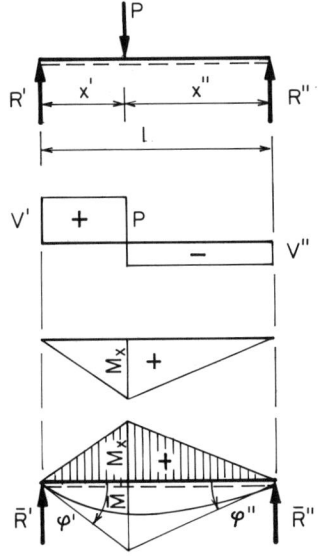

Figure 1-4(a)

---

*Otto Mohr, Abhandlungen 3 Afl., Berlin, 1928.

$$\Sigma Z = 0: \quad R' + R'' = P$$

$$\Sigma M = 0: \quad R'L - Px'$$

$$R' = \frac{Px''}{L} \qquad R'' = \frac{Px'}{L}$$

$$V' = R' \qquad V'' = R'' \qquad (1\text{-}14)$$

$$M_x = R'x' \text{ or } R''x''$$

$$EI\varphi' = \overline{R}' = \frac{M(L + x'')}{6}$$

$$EI\varphi'' = \overline{R}'' = \frac{M(L + x')}{6}$$

$$EI\delta = \overline{M} = \frac{Mx'x''}{3}$$

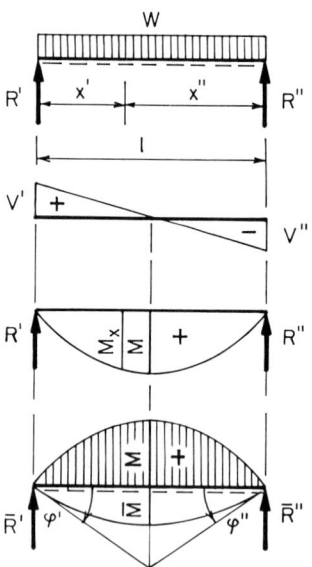

Figure 1-4(b)

$$\Sigma Z = 0: \quad R' + R'' - wL = 0$$

$$\Sigma M = 0: \quad R'L - \frac{wL^2}{2} = 0 \qquad (1\text{-}15)$$

Simple Beams

$$R' = R'' = \frac{wL}{2} = V' = V''$$

$$V_x = R' - wx$$

$$M_{max} = R' \frac{L}{2} - \frac{wL}{2} \frac{L}{4} = \frac{1}{8} wL^2$$

$$M_x = 4M_{max} \xi' \xi'' \qquad \xi' = \frac{x}{L} \qquad \xi'' = (1 - \xi')$$

$$EI\varphi' = EI\varphi'' = \frac{2}{3} \frac{wL^2}{8} \frac{L}{2} = \frac{wL^3}{24} = \overline{R}' = \overline{R}''$$

$$EI\delta_{max} = \overline{M} = \frac{5}{384} wL^4$$

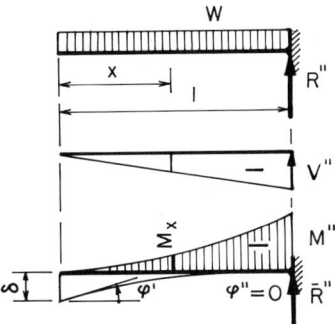

Figure 1-4(c)

$$\Sigma Z = 0: \qquad R'' - wL = 0, \qquad R' = 0$$

$$R'' = wL = V''$$

$$\Sigma M = 0: \qquad M'' - \frac{wL^2}{2} = 0 \tag{1-16}$$

$$M_x = -\frac{wx^2}{2}$$

$$EI\varphi' = \frac{wL^3}{6} \qquad EI\varphi'' = 0$$

$$EI\delta_{x=0} = \frac{wL^4}{8} = \overline{M}$$

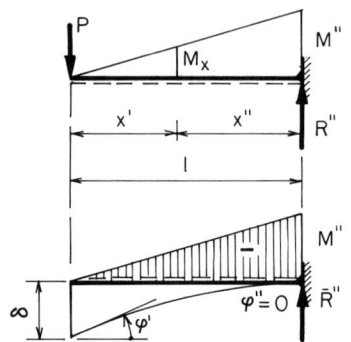

Figure 1-4(d)

$$\Sigma Z = 0: \quad R'' - P = 0, \quad R' = 0$$

$$\Sigma M = 0: \quad M'' = -PL$$

$$M_x = -P\,x \qquad (1\text{-}17)$$

$$EI\varphi'' = 0$$

$$EI\varphi' = \overline{R}'' = \frac{PL^2}{2}$$

$$EI\delta_{x=0} = \overline{M}'' = \frac{PL^3}{3}$$

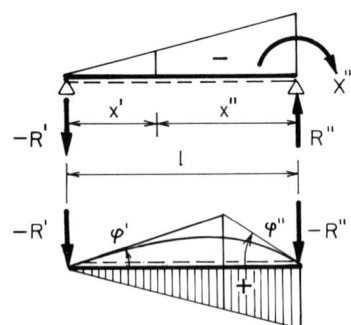

Figure 1-4(e)

$$\Sigma Z = 0: \quad R' + R'' = 0$$

$$R'' = -R'$$

$$\Sigma M = 0: \quad R'L = -X''$$

$$R' = -\frac{X''}{L}$$

Simple Beams

$$M_x = R'x \tag{1-18}$$

$$EI\varphi' = \overline{R}' = -\frac{X''L}{6}$$

$$EI\varphi'' = \overline{R}'' = -\frac{X''L}{3}$$

$$EI\delta_{\max} = \overline{M} = -\frac{X''}{6}\left(L - \frac{x}{L}\right)x$$

$$x \cong 0.58\,L$$

2

Statically Indeterminate Beams

METHOD OF ANALYSIS

Cases in which the three equations of equilibrium (Eq. 1-4) are not enough to determine the external reactions (R) and moments (M) are called *statically indeterminate systems*. To determine the redundants (X), the system is "cut" over the supports and reduced to a statically determined "principal system." In the principal system, the redundants (X', X'') act as the "cut faces." The required additional equations to the three equations of equilibrium are determined from the continuity requirements—i.e., the end rotations $EI\varphi$ of the joining members must be equal and their sum must be zero.

$$\varphi'' + \varphi' = 0 \qquad (2\text{-}1)$$

Such a "principal system" of a continuous beam, consisting of simple beams, is shown in Fig. 2-1.

The simple beams subjected to external loading ($w_D \cdot w_L$) deform and the ends suffer rotations ($EI\varphi$). The magnitude of the redundants X' and X'', acting in the principal system, must be such that reverse-end rotations ($EI\varphi$) caused by redundants (X', X'') will balance the ones due to the external loading—to satisfy the Eq. 2-1.

To determine the end rotation (φ), the loading, dimensions, and cross-sectional coefficients (Eqs. 1-7 to 1-12, Fig. 1-2a–e) must be known.

Method of Analysis

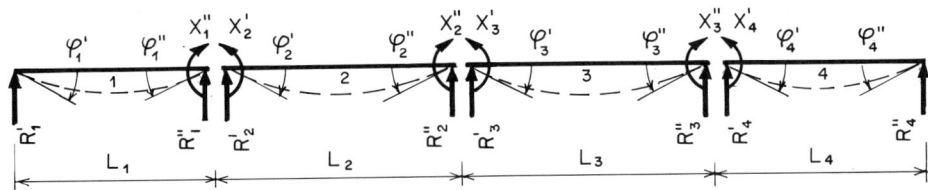

Figure 2-1

Zero-Point Method

A rather simple way to determine approximately the moments (X) and rotations (φ) for preliminary design and estimation is the *zero-point method*. Under the load, a beam deflects—and at supports, the deflections (δ) are zero. In locations at which the deflection curvature changes from downward to upward, the deflection line is straight; thus, the radius (r) is infinite, so the moment is $M = 0$. Such a point is called the inflection point (i.p.) or zero point. Between zero points, the beams act as simple beams and cantilevers. The reactions of simple beams are applied as concentrated loads for cantilevers.

The end-span moments often are critical, so that for dimensions only the end spans have to be computed. Thus, the support moment is:

$$R_{10} = \frac{wL_0}{2}, \qquad L_0 = L_1 - a''_1 L \tag{2-2}$$

$$X''_1 = -\left(R_{10}a''_1 + \frac{w \cdot a''^2_1}{2} = X'_2 = X_1\right)$$

and the midspan moment (M_m)

$$M_m = \frac{1}{8} wL_1^2 - \frac{X_1}{2}$$

wherein a''_1 is the zero point distance from the support. As guidance to estimate, the a-values may serve the one-end-fixed (Fig. 2-2) and fixed-end beams (Fig. 2-3).

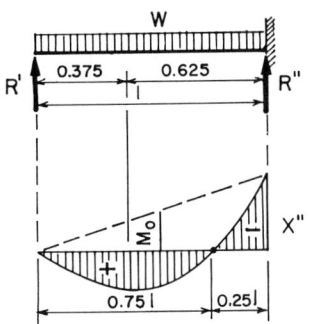

Figure 2-2

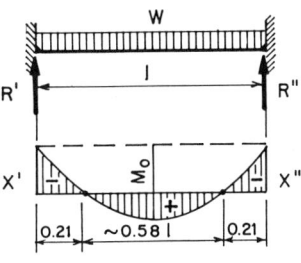

Figure 2-3

One-end fixed beam (Fig. 2-2) subjected to uniformly distributed load:

$$w = 1.0 \text{ t/m}, \ a_1' = 0, \ a_1'' = 0.25 \ L_1, \ L_0 = L_1 - a_1'' L_1 = 0.75 \ L_1$$

Fixed-end beam (Fig. 2-3):

$$w = 1.0 \text{ t/m}, \ a_1' = a_1'' \quad L_0 = L_1 - 2a_1 L_1 = 0.58 \ L_1$$

For continuous beams loaded with relatively light live load, the moments may be determined for full load ($w = w_D + w_L$). Since the ends are not fully fixed, the a-values are slightly less than for the one-end fixed and fixed-end beams.

Example (Fig. 2-4)

End span: $\quad w = 1.0 \text{ t/m}, \ L_1 = 10.00 \text{ m}, \ L_2 = 10.00 \text{ m}$

$$a_1'' \sim 0.21 \cdot L_1 = 2.10 \text{ m} \quad L_o = 10.00 - 2.10 = 7.90 \text{ m}$$

$$R_{10} = \frac{wL_o}{2} = \frac{1.0 \cdot 7.60}{2} = 3.95 \text{ t}$$

$$X_1'' = -\left(R_{10} a_1'' + \frac{w a_1''^2}{2}\right) = -\left(3.95 \cdot 2.10 + \frac{1.0 \cdot 2.10^2}{2}\right)$$

$$= -10.56 \text{ tm}$$

$$M_m = \frac{1}{8} wL^2 - \frac{X_1''}{2} = \frac{1}{8} 1.0 \cdot 10.00^2 - \frac{10.56}{2} = 7.20 \text{ tm}$$

$$R_0 = \frac{wL_1}{2} - \frac{X_1''}{L} = \frac{1.0 \cdot 10.00}{2} - \frac{10.56}{10.00} \cong 4.00 \text{ t}$$

$$R_1'' = \frac{1.0 \cdot 10.00}{2} + \frac{10.56}{10.00} \cong 6.00 \text{ t} = V_1''$$

$$R_2' \cong \frac{wL_2}{2} = \frac{1.0 \cdot 10.00}{2} = 5.00 \text{ t} \sim V_2'$$

$$R_1 = R_1'' + R_2' = 6.00 + 5.00 = 11.00 \text{ t}$$

Span 2: Because a_2' and a_2'' vary, it is rather difficult to estimate their values, therefore, the support moment X_2 may be assumed as fixed-end moment. Thus,

$$X_2 = \frac{1}{12} wL_2^2 \quad w = 1.0 \text{ t/m} \tag{2-3}$$

$$M_2 = \frac{1}{8} wL_2^2 - \frac{X_1 + X_2}{2}$$

The moments, reactions, and shears are illustrated in Fig. 2-4.

Method of Analysis

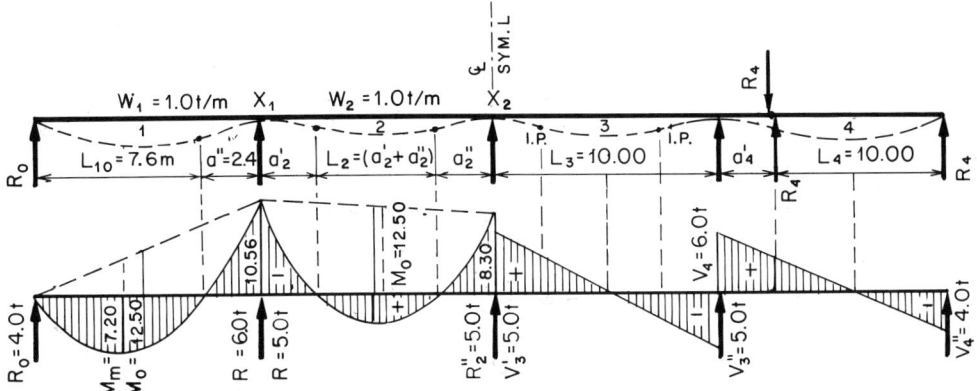

Figure 2-4

Example

Since depth (d_o) of a beam depends on loading, quality of concrete, and allowable vibration, it may vary from $L/d_o = 15$ to 20; b_o depends on the amount of steel, and it may vary from 20 to 30 cm. Assume $d_o = 10.00/17.5 \sim 0.60$ m, $d = d_o - d' = 60 - 4 = 56$ cm, $b_o = 25$ cm, the depth (t) of slab 12 cm, and spacing of beam 2.50 m. The dead load is $w_D = 0.84$ t/m. Live load is commonly given by code. For apartments and office buildings, $w_L = 250$ kg/m. Thus, the total load

$$w' = 0.84 + 2.50 \cdot 0.250 = 1.47 \text{ t/m} + \text{partitions}$$

$$w = \sim 1.60 \text{ t/m (multiplier)}$$

The support moment is critical. It is

$$X_1 = -10.56 \cdot 1.60 = -16.90 \text{ tm}$$

In accordance with Eqs. 1-5 and 1-6:

Support: $\quad T = -C = \dfrac{X_1}{z_o} = \dfrac{16.90}{0.48} \cong 35 \, t, \quad (z_0 \sim 0.86 \, d)$

$$A_s = \frac{T}{\sigma_s} = \frac{35}{1.400} = 25 \text{ cm}^2, \, (5 - {}^\#8)$$

$$V''_1 = \frac{V''_1}{b_o z_o} = \frac{1.6 \cdot 6 \cdot 1000}{25 \cdot 48} \cong 8 \text{ kg/cm}^2 < 10 \text{ kg/cm}^2$$

Concrete strength at the time of stripping of the forms is about 300 kg/cm² for 350 kg/cm² cylinder strength and modulus of elasticity in accordance with ROS:

$$E_c = 370000 \text{ kg/cm}^2, \, n = \frac{E_s}{E_c} = \frac{2100000}{370000} \sim 5.7$$

$$\sigma_c = 0.40\sigma_{uL} \cong 0.120 \text{ t/cm}^2, \ (\sigma_{uL} = 0.300 \text{ t/cm}^2)$$

$$k = \frac{n\sigma_c}{n\sigma_c + \sigma_s} = \frac{5.7 \cdot 0.120}{5.7 \cdot 0.120 + 1.400} \cong 0.33$$

$$\sigma_c = \frac{2T}{b_0 k d} = \frac{2 \cdot 35}{25 \cdot 0.33 \cdot 56} = 0.150 \text{ t/cm}^2$$

Span:

$$M_m = 1.60 \cdot 7.2 \cong 11.50 \text{ tm}$$

$$z_0 = d_0 - \frac{t}{2} = 56 - \frac{12}{2} = 50 \text{ cm}$$

$$T = \frac{M_{m1}}{z_0} = \frac{11.50}{0.50} = 23 \text{ t}, \ (4\text{-}^{\#}8)$$

The sections are shown in Figs. 2-5 and 2-6.

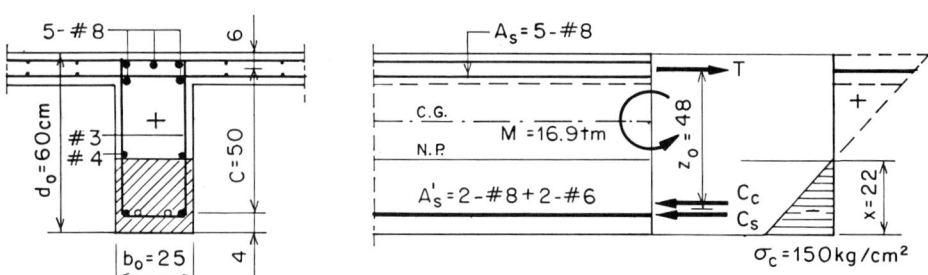

Figure 2-5

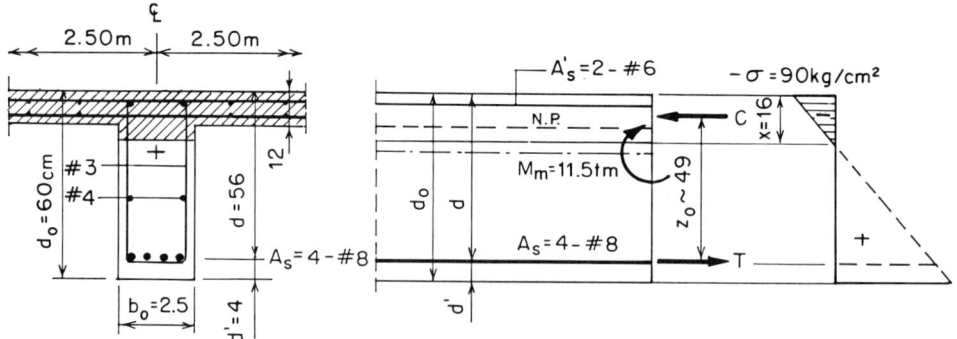

Figure 2-6

Since the concrete stresses at support exceed what is allowable, compressive reinforcing is required. As steel is not subjected to shrinkage and creep, it additionally reduces the time deflection ($t = n$).

Method of Analysis

Such preliminary computations can be accomplished by the use of a pocket calculator within three to four hours.

For final structural analysis, the zero-point method is inadequate, especially where length of spans (L) and loading (w) differ. In the following section, a rather simple method will be discussed.

Loading Factor Method

In practice, the Cross or Moment Distribution methods are commonly used to determine the moments of continuous beams. These methods are time-requiring and do not create the feeling of how the structure really works. On the contrary, the load factor method based on the three-moments equation is much simpler, requiring less time and clearly showing how the structure behaves under load.

In accordance with Otto Mohr, the end rotation for simply supported beams is obtained by loading a beam with the external moment (M). The reactions ($\overline{R}$) of such moment loading are the end rotation $EI\varphi' = \overline{R}'$ and $EI\varphi'' = \overline{R}''$. For symmetric loading, $EI\varphi' = EI\varphi''$.

The loading factors ($\overline{L}$) are obtained by multiplying the reactions $\overline{R}_n$ by $\dfrac{6}{L_n}$; thus:

$$\overline{L}'_n = \frac{6}{L_n} \overline{R}'_n$$

$$\overline{L}''_n = \frac{6}{L_n} \overline{R}''_n \tag{2-4}$$

Loading factors. For common loadings, the $\overline{R}$ and $\overline{L}$-values are computed as in the following equation (see Chapter 1).

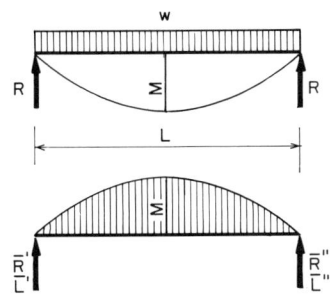

Figure 2-7

$$M_{max} = \frac{1}{8} wL^2$$

$$\overline{R}' = \frac{2}{3} \frac{L_n}{2} M_{max} = \frac{1}{24} wL^3 = \overline{R}'' \tag{2-5}$$

$$\bar{L}' = \bar{R}' \cdot \frac{6}{L_n} = \frac{1}{24} wL_n^3 \cdot \frac{6}{L_n}$$

$$= \frac{1}{4} wL_n^2 = 2M_{\max}$$

$$= \bar{L}''$$

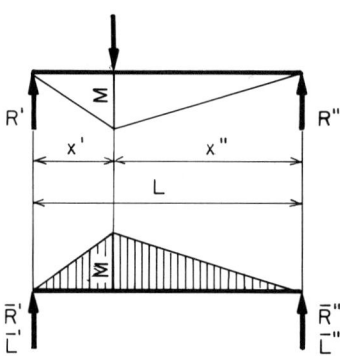

Figure 2-8

$$R' = \frac{Px''}{L}, \qquad R'' = \frac{Px'}{L}$$

$$M' = R'x' = \frac{Px''}{L} x'$$

$$M'' = R''x'' = \frac{Px'}{L} x''$$

$$\bar{R}' = \frac{M_{\max}(L + x'')}{6}$$

$$\bar{R}'' = \frac{M_{\max}(L + x')}{6} \qquad (2\text{-}6)$$

$$\bar{L}' = \frac{Px'x''}{L^2}(L + x'')$$

$$\bar{L}'' = \frac{Px'x''}{L^2}(L + x')$$

$$x' = x'': \qquad \bar{L}' = \bar{L}'' = \frac{3PL}{8}$$

Method of Analysis

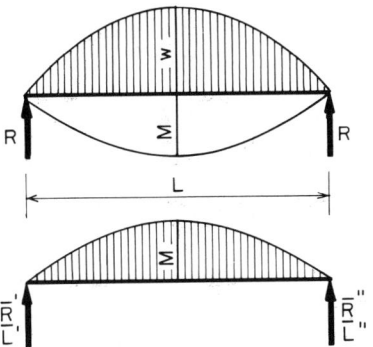

Figure 2-9

$$M = \frac{1}{10} wL^2$$

$$\overline{R}' = \frac{2}{3}\frac{L}{2} M = \frac{1}{30} wL^3 = \overline{R}'' \qquad (2\text{-}7)$$

$$\overline{L}' = \frac{1}{30} \cdot \frac{6}{L} wL^3 = \frac{1}{5} wL^2 = \overline{L}''$$

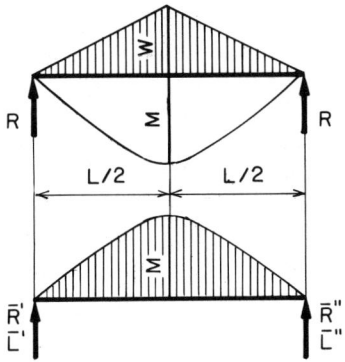

Figure 2-10

$$R' = R'' = \frac{1}{4} wL$$

$$M = \frac{2}{3} R' \frac{L}{2} = \frac{1}{12} wL^2 \qquad (2\text{-}8)$$

$$\bar{R}' = \frac{5}{196} wL^3 = \bar{R}''$$

$$\bar{L}' = \bar{R}' \frac{6}{L_n} = \frac{5}{32} wL^2 = \bar{L}''$$

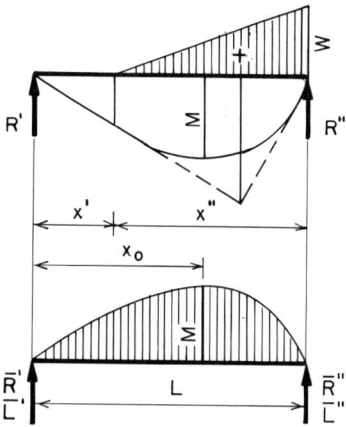

Figure 2-11

$$R' = \frac{x''^2}{6L} \, w \qquad R'' = \frac{x''}{6L}\left(3 - \frac{x''}{L}\right)$$

$$M_{\max} = \frac{x''^2}{6L}\left(x' + \frac{2}{3} x'' \sqrt{\frac{x''}{3L}}\right)$$

$$x_0 = x' + x'' \sqrt{\frac{x''}{3L}}$$

$$\bar{R}' = \frac{x''^2}{360L}\left(10L^2 - 3x''^2\right) w \qquad (2\text{-}9)$$

$$\bar{R}'' = \frac{x''^2}{360L}(20L^2 - 15x''L + 3x''^2)\, w$$

$$\bar{L}' = \bar{R}' \frac{6}{L} = \frac{x''^2}{60L^2}(10L^2 - 3x''^2)\, w$$

$$\bar{L}'' = \bar{R}'' \frac{6}{L} = \frac{x''^2}{60L^2}(10L^2 - 15x''L + 3x''^2)\, w$$

Method of Analysis

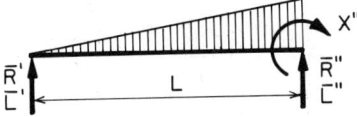

Figure 2-12

$$R' = -\frac{X''}{L}, \qquad R'' = \frac{X''}{L}$$

$$\overline{R}' = \frac{X''}{6} L, \qquad \overline{R}'' = \frac{X''}{3} L \qquad (2\text{-}10)$$

$$\overline{L}' = \overline{R}' \frac{6}{L} = X''$$

$$\overline{L}'' = \overline{R}'' \frac{6}{L} = 2X''$$

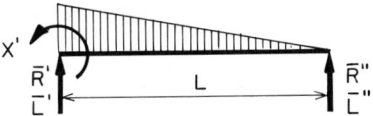

Figure 2-13

$$\overline{L}' = 2X''$$
$$\overline{L}'' = X'' \qquad (2\text{-}11)$$

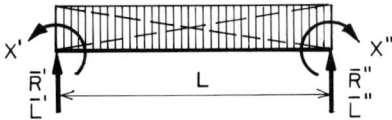

Figure 2-14

$$\overline{L}' = 2X' + X''$$
$$\overline{L}'' = 2X'' + X' \qquad (2\text{-}12)$$

Beam end rotations. The end rotations at supports $EI\varphi_n''$ and $EI\varphi_{n+1}'$ for a continuous beam, shown in Fig. 2-15, are:

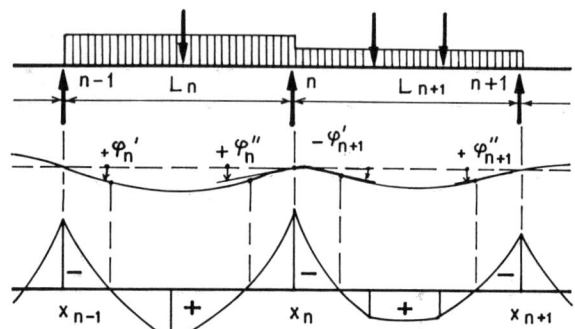

Figure 2-15

$$EI\varphi_n'' = L_n \left(\frac{1}{3} X_n + \frac{1}{6} X_{n-1} + \overline{R}_n'' \right)$$

$$EI\varphi_{n+1}' = L_{n+1} \left(\frac{1}{3} X_n + \frac{1}{6} X_{n+1} + \overline{R}_n' \right)$$

(2-13)

Substituting the $\overline{R}$-values by loading factor $\overline{L}$, we obtain

$$\overline{L}'' = \frac{6\overline{R}''}{L}, \qquad \overline{R}'' = \frac{L}{6} \overline{L}''$$

$$\overline{L}' = \frac{6\overline{R}'}{L}, \qquad \overline{R}' = \frac{L}{6} \overline{L}'$$

$$EI\varphi_n'' = \frac{L_n}{6} (2X_n + X_{n-1} + \overline{L}_n'')$$

$$EI\varphi_{n+1} = \frac{L_{n+1}}{6} (2X_n + X_{n+1} + \overline{L}_{n+1}')$$

(2-14)

Continuity requires:

$$\varphi_n'' + \varphi_n' = 0 \qquad (2\text{-}15)$$

Introducing Eq. 2-14 into Eq. 2-15, we obtain for support n

$$L_n (2X_n + X_{n-1} + \overline{L}_n'') + L_{n+1} (2X_n + X_{n+1} + \overline{L}_{n+1}') = 0 \qquad (2\text{-}16)$$

Such an equation can be written for each support. The redundants (X) are obtained by solving these simultaneous linear equations.

Span-by-Span Loading

For simplification of the analysis, span-by-span loading is suggested. The final reactions (R), moments (M), and redundants (X) will then be obtained by superposition (assumption 3). In this case, the loading factors ($\overline{L}$) for the unloaded spans are zero. The redundants X'' and X' in loaded span n are active.

Method of Analysis

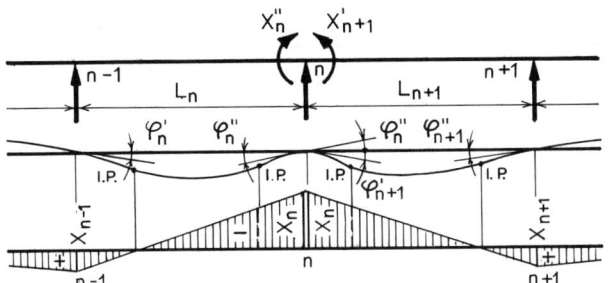

Figure 2-16

Carryovers. Denoting the ratio of support moments by c and starting from the left, we find the c'-values as follows:

$$c'_{n-1} = -\frac{X_{n-1}}{X_n}, \qquad -X_{n-1} = c_{n-1}X_n$$

$$c'_{n+1} = -\frac{X_n}{X_{n+1}}, \qquad -X_{n+1} = \frac{X_n}{c'_{n+1}} \tag{2-17}$$

Introducing the $-X_{n-1}$ and $-X_{n+1}$ values into Eq. 2-16, we obtain

$$c'_{n+1} = \frac{L_{n+1}}{2(L_n + L_{n+1}) - c'_n L_n} \tag{2-18}$$

Similarly, starting from the right, we find the c''-values

$$c''_{n-1} = -\frac{X_n}{X_{n-1}}, \qquad -X_{n-1} = \frac{X_n}{c''_{n-1}}$$

$$c''_{n+1} = -\frac{X_{n+1}}{X_n}, \qquad -X_{n+1} = c''_{n+1}X_n \tag{2-19}$$

$$c''_{n-1} = \frac{L_n}{2(L_n + L_{n+1}) - c''_{n+1} L_{n+1}} \tag{2-20}$$

The relationship between the carryover (c) and distance (a) of the zero points from the supports are

$$c' = \frac{a'}{L - a'} \qquad a' = \frac{c'L}{1 + c'}$$

$$c'' = \frac{a''}{L - a''} \qquad a'' = \frac{c'L}{1 + c''} \tag{2-21}$$

The a- and c-values for each span are indicated in Fig. 2-17.

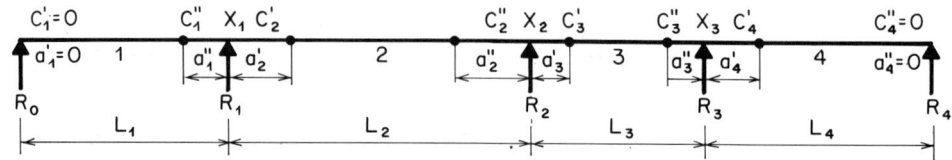

Figure 2-17

For example, the $c_1'c''$- and $a_1'a''$-values for a continuous beam (Fig. 2-17) will be computed as follows: $L_1 = 4.5$ m, $L_2 = 6.5$ m, $L_3 = 4.0$ m, $L_4 = 5.0$ m.

0 to 4:

$$c_2' = \frac{L_2}{2(L_1 + L_2) - c_1'L_1} = \frac{6.5}{2(4.5 + 6.5) - 0 \cdot 4.5} = 0.295, \quad a_2' = 1.48 \text{ m}$$

$$c_3' = \frac{L_3}{2(L_2 + L_3) - c_2'L_2} = \frac{4.0}{2(6.5 + 4.0) - 0.295 \cdot 6.5} = 0.209, \quad a_3' = 0.69 \text{ m}$$

$$c_4' = \frac{L_4}{2(L_3 + L_4) - c_3'L_3} = \frac{5.0}{2(4.0 + 5.0) - 0.209 \cdot 4.0} = 0.291, \quad a_4' = 1.13 \text{ m}$$

4 to 0:

$$c_3'' = \frac{L_3}{2(L_4 + L_3) - c_4''L_4} = \frac{4.0}{2(5.0 + 4.0) - 0 \cdot 5.0} = 0.222, \quad a_3'' = 0.73 \text{ m}$$

$$c_2'' = \frac{L_2}{2(L_3 + L_2) - c_3''L_3} = \frac{6.5}{2(4.0 + 6.5) - 0.222 \cdot 4.0} = 0.323, \quad a_2'' = 1.59 \text{ m}$$

$$c_1'' = \frac{L_1}{2(L_2 + L_1) - c_2''L_2} = \frac{4.5}{2(6.5 + 4.5) - 0.323 \cdot 6.5} = 0.226, \quad a_1'' = 0.83 \text{ m}$$

If the beam ends are elastically restrained, the a_1'-value may be taken 0.75 of the fully restrained a-value. Thus, by Eq. 2-21:

$$a_1' = 0.75 \, aL_1 = 0.75 \cdot 0.21 \, L_1 \sim 0.16 \, L_1 \text{ m}$$

$$c_1' = \frac{a_1''}{L_1 - a_1''} = \frac{0.16 \, L_1}{L - 0.16L_1} = \frac{0.16}{1 - 0.16} \sim 0.19$$

SUPPORT MOMENTS

By Eq. 2-16, the equations for supports n'' and n' for the loaded span n are (Fig. 2-18):

$$\begin{aligned} -(X_n'/c_n' + X_n'') + L_n' = 0 \\ -(X_n''/c_n'' + X_n') + L_n'' = 0 \end{aligned} \quad (2\text{-}22)$$

Support Moments

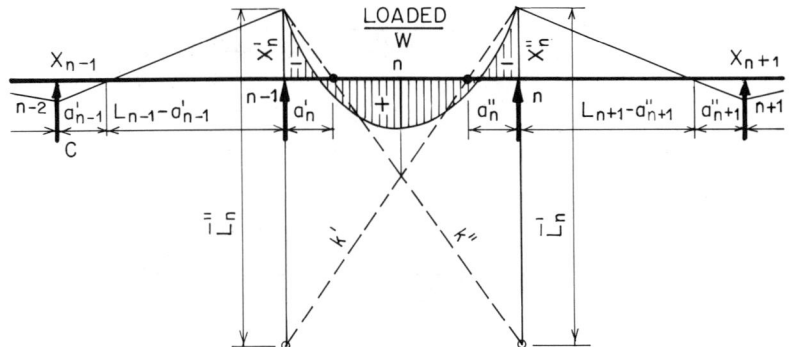

Figure 2-18

By solving Eq. 2-22 with respect to X' and X'', we obtain:

$$X'_n = -c'_n \frac{\overline{L}'_n - c''_n \overline{L}''}{1 - c'_n c''_n}$$

$$X''_n = -c''_n \frac{\overline{L}''_n - c'_n \overline{L}'_n}{1 - c'_n c''_n}$$

(2-23)

For symmetric loading of the span, $\overline{L}' = \overline{L}''$. Thus,

$$X'_n = -c'_n \overline{L}'_n \frac{1 - c''_n}{1 - c'_n c''_n}$$

$$X''_n = -c''_n \overline{L}''_n \frac{1 - c'_n}{1 - c'_n c''_n}$$

(2-24)

For simply supported ends of the beam, $c'_1 = c''_n = 0$. Thus,

$$X'_n = -c'_n \overline{L}'_n$$

$$X''_n = -c''_n \overline{L}''_n$$

(2-25)

For checking the correctness of computations, the k' and k''-lines have to pass the zero points a' and a''.

Example

Uniform Loading

One-span beams. In accordance with Eqs. 2-5, 2-10, and 2-16 (Figs. 2-19 and 2-20):

$$\varphi'' = 0$$

$$\overline{L}' = \overline{L}'' = \overline{L} = \frac{wL^2}{4}$$

28 Chap. 2 Statically Indeterminate Beams

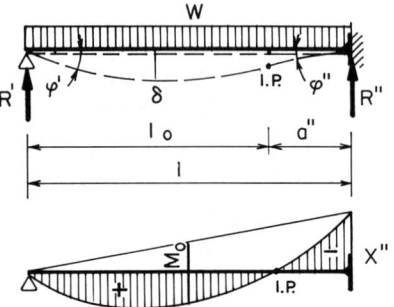

Figure 2-19

$$\overline{L}_x = 2X''$$

$$2X'' + \overline{L}'' = 0$$

$$X'' = -\frac{\overline{L}''}{2} = -\frac{wL^2}{2 \cdot 4} = -\frac{1}{8} wL^2$$

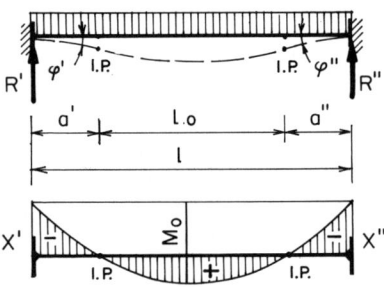

Figure 2-20

$$\varphi' = \varphi'' = 0$$

$$\varphi' = 2X' + X'' + \overline{L} = 0$$

$$\varphi'' = 2X'' + X' + \overline{L} = 0$$

$$X' = X'' = -\frac{\overline{L}}{3} = -\frac{wL^2}{3 \cdot 4} = -\frac{1}{12} wL^2$$

Two-span beams. See Fig. 2-21.

$$w_1 = 1.0 \qquad w_2 = 0 \qquad c_1' = 0$$

$$L_1 = L_2 = 1.0 \qquad c_1'' = c_2' = \frac{1.0}{2(1.0 + 1.0)} = 0.25$$

Support Moments

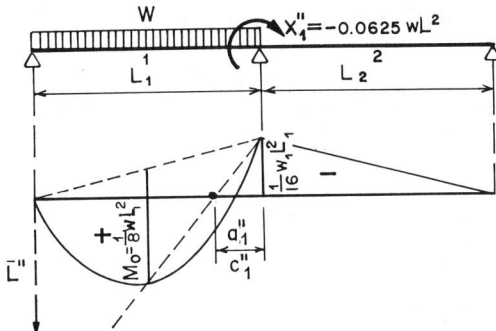

Figure 2-21

$$X''_1 = -c''_1 \bar{L}_1 = -0.25 \cdot \frac{wL_1^2}{4}$$

$$= -0.0625 \, wL_1^2 = X'_2$$

$$X_1 = X''_1 + X'_2 = -0.125 \, wL^2$$

$$L_2 = 0.8L_1 = 0.80 \cdot 1.0$$

$$c''_1 = \frac{1.0}{2(1 + 0.80)} = 0.278$$

$$c'_2 = \frac{0.80}{2(1 + 0.80)} = 0.222$$

$$X''_1 = -c''_1 \bar{L}_1 = -0.278 \, \frac{wL_1^2}{4} = -0.069 \, wL_1^2 \text{ tm}$$

$$X'_2 = -c'_2 \bar{L}_2 = -0.222 \, \frac{wL_2^2}{4} = -0.056 \, wL_2^2 \text{ tm}$$

$$X_1 = X''_1 + X_2 = -(0.069 \, wL_1^2 + 0.056 \, wL_2^2) \text{ tm}$$

$$M_1 = M_{01} - \frac{X_1}{2} \qquad M_2 = M_{02} - \frac{X_1}{2} \qquad M_0 = \frac{1}{8} wL_2^2$$

$$R'_1 = R_{01} - \frac{X_1}{L_1} \qquad R''_1 = R_{01} + \frac{X_1}{2} = V''_1 \qquad R_{01} = \frac{wL_1}{2}$$

$$R''_2 = R_{02} - \frac{X_1}{2} \qquad R'_2 = R_{02} + \frac{X_1}{2} \qquad R_{02} = \frac{wL_2}{2}$$

$$R_1 = R''_1 + R'_2$$

Three-span beams.

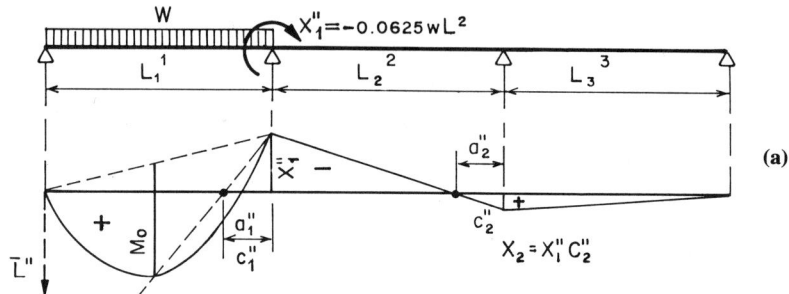

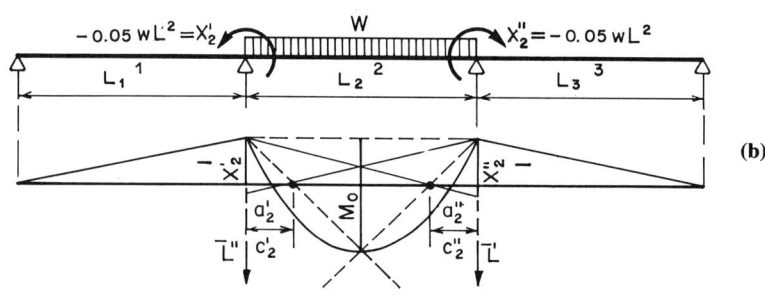

Figure 2-22

$$w_1 = 1.0 \text{ t/m} \qquad w_2 = w_3 = 0 \qquad L_1 = L_3 = 1.0 \qquad \bar{L}_n = \frac{wL_n}{4}$$

$$L_2 = 1.2 \, L_1 = 1.2$$

$$c_1'' = \frac{L_1}{2(L_1 + L_2)} = \frac{1.0}{2(1.0 + 1.2)} = 0.227 = c_3'$$

$$c_2' = c_2'' = \frac{L_2}{2(L_1 + L_2)} = \frac{1.2}{2(1.0 + 1.2)} = 0.273$$

$$X_1'' = X_3' = -c_1'' \bar{L}_1 = -0.227 \cdot \frac{w_1 L_1^2}{4} = -0.0568 \, wL_1^2$$

$$X_2 = X_1 = c_2'' X_1'' = 0.273 \cdot 0.0568 \, wL_1^2 = 0.0155 \, wL_1^2$$

$$w_2 = 1.0 \text{ t/m} \qquad w_1 = w_3 = 0$$

$$X_2' = X_2'' = -c_2' \bar{L}_2' \frac{1 - c_2''}{1 - c_2' c_2''} = -0.273 \, \bar{L}_2 \frac{1 - 0.273}{1 - 0.273 \cdot 0.273} \, wL_2^2$$

$$= -0.0536 \, wL_2^2$$

Support Moments

$$\Sigma X_1 = X_2 = X_1'' + X_2' + X_1 = -[0.0568 \cdot L_1^2 - (0.0536 - 0.0155)L_2^2]w$$
$$= -(0.0568 L_1^2 + 0.0381 L_2^2)w$$

$$L_1 = L_2 = L_3 = 1.0: \qquad c_1'' = c_2' = c_2'' = c_3' = 0.25$$

$$X_1 = X_2 = -(0.0625 + 0.0500)wL^2 = -0.1125\ wL^2$$

$$M_1 = M_{01} - \frac{X_1}{2} = M_3 \qquad M_2 = M_{02} - \frac{X_1 + X_2}{2}$$

$$R_0 = \frac{1.0 \cdot 1.0}{2} wL$$

$$R_1'' = R_{01} + \frac{0.0625\ wL^2}{L_1} = \left(0.500 + \frac{0.0625}{1.0}\right)$$

$$= 0.5625\ wL$$

$$R_1 = (0.5625 + 0.500)\ wL = 1.0625\ wL$$

Four-span beams.

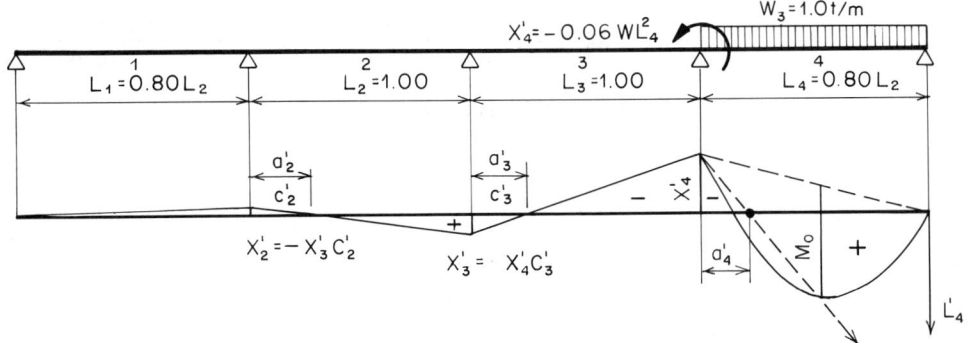

Figure 2-23

$$L_2 = L_3 = 1.0 \qquad L_1 = L_4 = 0.80\ L_2 = 0.80$$

$$c_1' = c_4'' = 0$$

$$c_2' = c_3'' = \frac{0.80}{2(1.0 + 0.80)} = 0.222$$

$$c_3' = c_2'' = \frac{L_3}{2(L_2 + L_3) - c_2'L_2} = \frac{1.0}{2(1.0 + 1.0) - 0.222 \cdot 1.0} = 0.265$$

$$c'_4 = c''_1 = \frac{L_4}{2(L_3 + L_4) - c'_3 L_3} = \frac{0.80}{2(1.0 + 0.80) - 0.265 \cdot 1.0} = 0.240$$

$$w_1 = 1.0 \quad X''_1 = X'_4 = -c''_1 \overline{L}_1 = -0.240 \frac{wL_1^2}{4} = -0.060 wL_1^2$$

$$X_2 = +X''_1 c''_2 = 0.060 \cdot 0.265 \ wL_1^2 = +0.0159 \ wL_1^2$$

$$X_3 = -X_2 c''_3 = -0.0159 \cdot 0.222 \ wL_1^2 = -0.0035 \ wL_1^2$$

The moments diagrams are illustrated in Figs. 2-23 and 2-24.

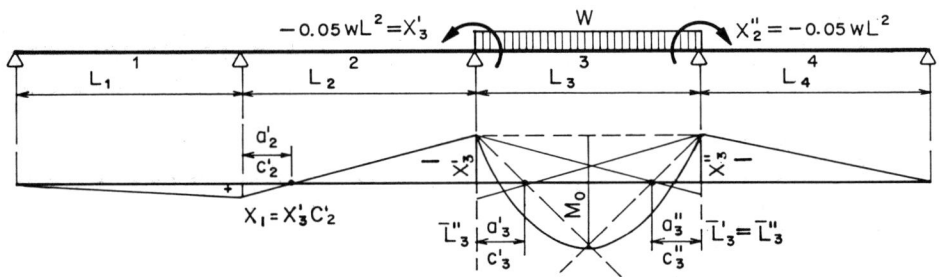

Figure 2-24

$$w_2 = 1.0 \quad X'_2 = -c'_2 \frac{1 - c''_2}{1 - c'_2 c''_2} \overline{L}_2 = -0.222 \frac{1 - 0.265}{1 - 0.222 \cdot 0.265} \overline{L}_2$$

$$= -0.0434 \ wL_2^2 = X''_3$$

$$X''_2 = X'_3 = -c''_2 \frac{1 - c'_2}{1 - c'_2 c''_2} \overline{L}_2 = -0.265 \frac{1 - 0.222}{1 - 0.222 \cdot 0.265} \overline{L}_2 = -0.0548 \ wL_2^2$$

$$X_3 = X_1 = c''_3 X''_2 = +0.222 \cdot 0.0548 \ wL_2^2 = +0.0122 \ wL_2^2$$

$$\Sigma X_1 = \Sigma X_3 = -(X''_1 + X'_2 - X_3 + X_1)$$

$$= -(0.600 + 0.0035) \ wL_1^2 - (0.434 - 0.122) \ wL_2^2$$

$$= -(0.6035 \ L_1^2 + 0.312 \ L_2^2)w$$

$$M_1 = M_{01} - \frac{\Sigma X_1}{2} = M_4 \qquad M_2 = M_{02} - \Sigma X_1 = M_3$$

$$L_1 = L_2 = L_3 = L_4 \qquad c'_2 = c''_3 = 0.25 \qquad c'_3 = c''_2 = 0.267 = c''_1 = c'_4$$

$$X''_1 = -c''_1 \overline{L} = -0.267 \frac{wL^2}{4} = -0.0670 \ wL_1 = X'_4$$

Support Moments

$$X_2' = -c_2' \frac{1 - c_2''}{1 - c_2'c_2''} \bar{L} = -0.25 \frac{1 - 0.267}{1 - 0.25 \cdot 0.267} \bar{L} = -0.05\, wL^2$$

$$X_2'' = -c_2'' \frac{1 - c_2'}{1 - c_2'c_2''} \bar{L} = -0.267 \frac{1 - 0.25}{1 - 0.25 \cdot 0.267} \bar{L} = -0.054\, wL^2$$

$$c_1 = 0.25 \text{ (constant)}$$

$$X_1'' = -0.0625\, wL_1^2 \qquad \Delta X_1'' = 0.0670 - 0.0625 = 0.0045\, wL_1^2$$

$$X_2' = X_2'' = 0.050\, wL_2^2 \qquad \Delta X_2' \sim 0.0 \qquad \text{(Figs. 2-23 and 2-24)}.$$

Concentrated Loads $P = 1.0$

The carryovers (c', c'') are independent of loading; therefore in computation of moments due to concentrated loads P, only the $\bar{L}$-values change. Otherwise, the computations are the same as for uniform loading.

In accordance with Eq. 2-6, the $\bar{L}$-value for spacing of $0.333\, L$ is:

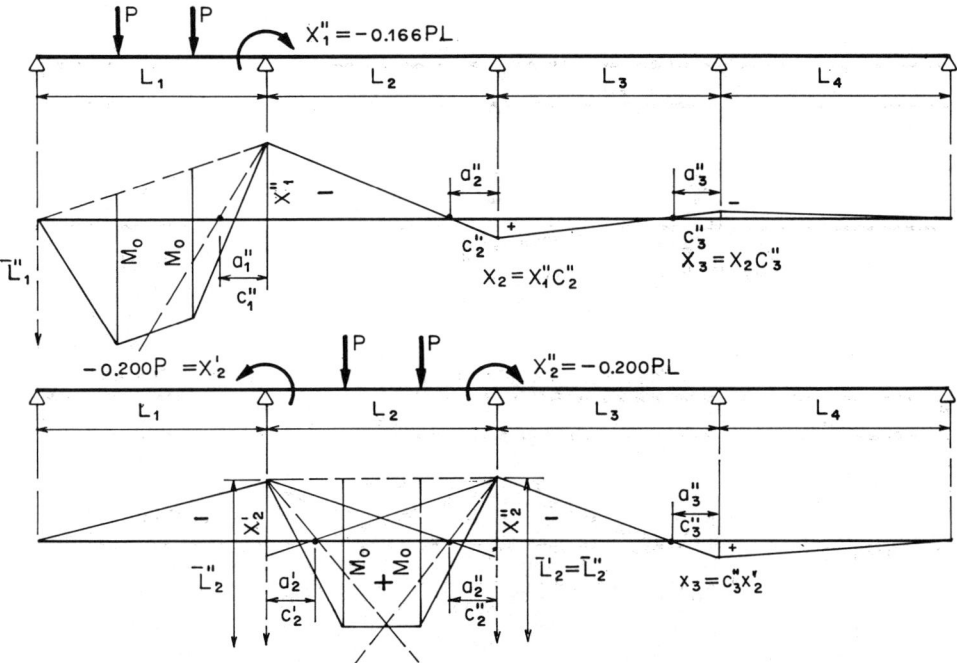

Figures 2-25 and 2-26

$$P_1 = P_2 = 1.0\text{ t} \qquad L = 1.0$$

$$P_1 = 1.0:\ x' = 0.333,\ x'' = 0.666$$

$$\overline{L}_1' = \frac{x'x''}{L^2}(1 + x'')L = 0.369$$

$$\overline{L}_1'' = \frac{x'x''}{L^2}(1 + x')L = 0.296$$

$$P_2 = 0: \quad \overline{L}_2' = \frac{x'x''}{L^2}(1 + x')L = 0.296 \qquad \overline{L}_2'' = 0.369$$

$$\Sigma\overline{L}' = \overline{L}_1' + \overline{L}_2'' = (0.369 + 0.296)\,PL = 0.665\,PL$$

$$C = 0.25 \quad \text{(constant)} \quad \text{(Figs. 2-25 and 2-26)}$$

$$X_1'' = X_4' = -c\overline{L} = -0.25 \cdot 0.665\,PL = -0.166\,PL$$

$$X_2' = X_3'' = -cL\frac{1-c}{1-c^2} = -0.25\frac{1-0.25}{1-0.25^2}PL = -0.200\,PL$$

$$X_2 = X_1''c = 0.166 \cdot 0.25 \cdot PL = +0.042\,PL$$

$$X_3 = X_2 c = -0.042 \cdot 0.25 = -0.011\,PL$$

$$X_1 = X_3 = -0.011\,PL$$

$$X_2'' = X_3' = -0.200\,PL$$

$$X_3 = X_2''c = +0.200 \cdot 0.25 = 0.050\,PL = X_1$$

$$\Sigma X_1 = \Sigma X_3 = -(X_1'' + X_2' + X_1 - X_2)\,PL$$
$$= -(0.166 + 0.200 + 0.011 - 0.050)\,PL = -0.327\,PL$$

$$\Sigma X_2 = -(X_2'' + X_3' - 2 \cdot X_2)\,PL = -(0.200 + 0.200 - 2 \cdot 0.042)\,PL$$
$$= -0.316\,PL$$

$$M_1 = M_4 = M_{10} - \frac{\Sigma X_1}{2} \qquad M_2 = M_3 = M_{20} - \frac{X_1 + X_2}{2}$$

CRITICAL LIVE LOAD LOCATIONS

Continuous beams in apartment and office buildings are generally subjected to uniformly distributed dead and live loads ($w = w_D + w_L$) for all spans. However, in utility buildings, such as warehouses, factories, and so on, live load can be rather heavy and may change from span to span. To obtain maximum moments, live load must be considered located, as illustrated in Fig. 2-27.

Loading 1 $M_{1,\max}, M_{3,\max}$

Loading 1 $M_{2,\min}, M_{4,\min}\ R_{0,\max}$

Loading 2 $X_{1,\max}, M_{4,\max}$

Critical Live Load Locations

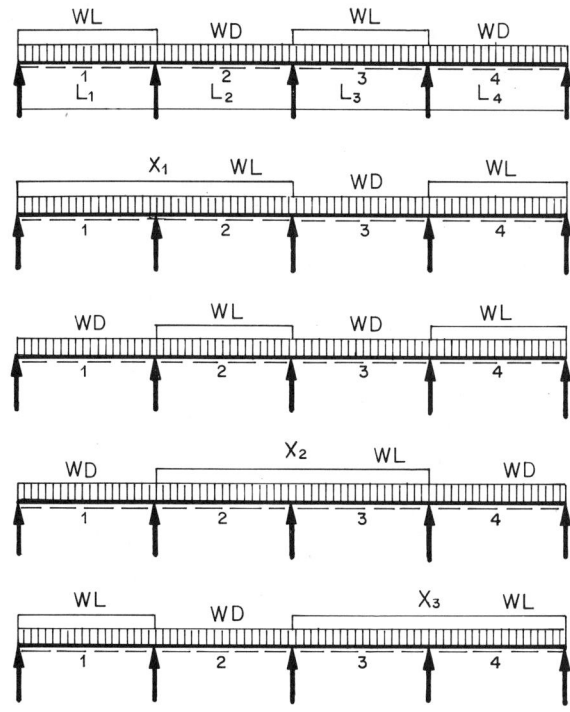

Figure 2-27

Loading 3 $M_{2,\max}$, $M_{4,\max}$

Loading 3 $M_{1,\min}$, $M_{3,\min}$ $R_{4,\max}$

Loading 4 $X_{2,\max}$ $R_{2,\max}$

Loading 5 $X_{3,\max}$ $R_{3,\max}$

As can be seen from the foregoing computations, carried out for equal spans ($c = 0.25$) and for span lengths with variation of less than 20 percent ($c = $ variable), the difference in redundants ($X'_n X''_n$) varies less than 5 percent. To obtain maximum-minimum moments (M_n) most easily, superposition of the span-by-span loading for constant $c = 0.25$ has been computed for $w_D = 1.0$ and $w_L = 1.0$ and presented in Table 1 for uniform load and in Table 2 for concentrated loads, in which only the outside spans and one intermediate span are considered.

However, if the difference in spans varies more than 20 percent, the c-values have to be determined by Eqs. 2-18 and 2-20, and the redundants $X'_n X''_n$ by Eqs. 2-23, 2-24, and 2-25.

In cases where difference in spans is less than 20 percent, the moment and reaction coefficients for uniform and concentrated loads are computed and given in Tables 1 and 2, included in the Appendix.

36 Chap. 2 Statically Indeterminate Beams

For illustration of the use of Tables 1 and 2, the maximum moments (X_n, M_n) of a four-span continuous beam under dead load (w_D, P_D) and moving live load (w_L) on spans (Fig. 2-27) will be computed in the following.

Example

$$L_1 = L_4 = 10.00 \text{ m} \qquad L_2 = L_3 = 12.00 \text{ m}$$

Dead load: $w_D = 1.0$ t/m

$$X_1 = [-(0.0625\,L_1^2 + 0.050\,L_2^2 - 0.0125 L_3^2 + 0.0039\,L_4^2)]w_D$$
$$= [-(6.25 + 7.20 - 1.80 + 0.39)]\,1.0 = -12.04 \text{ tm} = X_3$$

$$X_2 = [-(0.050\,L_2^2 + 0.050\,L_3^2) + (0.0156\,L_1^2 + 0.0156\,L_4^2)]w_D$$
$$= [-(7.20 + 7.20) + (1.56 + 1.56)1.0 = -11.28 \text{ tm} = X_3$$

$$M_{m1} = [(0.0950\,L_1^2 + 0.0050\,L_3^2) - (0.0200\,L_2^2 + 0.0016\,L_4^2)]w_D$$
$$= 6.48 \text{ tm} = M_{m4} = M_{10} - \frac{X_1}{2}$$

$$M_{m2} = [(0.0750\,L_2^2 + 0.0059\,L_4^2) - (0.0234\,L_1^2 + 0.0188\,L_3^2)]w_D$$
$$= 7.18 \text{ tm} = M_{m3} = M_{30} - \frac{X_1 + X_2}{2}$$

Live load: $w_L = 1.0$ t/m

$$X_{1,\max} = -(0.0625\,L_1^2 + 0.050\,L_2^2 + 0.0039\,L_4^2)w_L$$
$$= -(6.25 + 7.20 + 0.39)1.0 = -13.84 \text{ tm} = X_3$$

$$X_{2,\max} = -(0.050\,L_2^2 + 0.050\,L_3^2)w_L$$
$$= -(7.20 + 7.20)1.0 = -14.40 \text{ tm}$$

$$M_{1\max} = (0.095\,L_1^2 + 0.0050\,L_3^2)w_L = (9.50 + 0.72)1.0 = 10.22 \text{ tm} = M_{m4}$$

$$M_{2\max} = (0.075\,L_2^2 + 0.0059\,L_4^2)w_L = (10.80 + 0.59)1.0 = 11.39 \text{ tm} = M_{m3}$$

Summary: $\Sigma M_{\max} = M_D + M_L$

$$X_{1,\max} = -(12.04 + 13.84) = -25.88 \text{ tm} = X_{3,\max}$$

$$X_{2,\max} = -(11.28 + 14.40) = -25.68 \text{ tm}$$

$$M_{1,\max} = 6.48 + 10.22 = 16.70 \text{ tm} = M_{4,\max}$$

$$M_{2,\max} = 7.18 + 11.39 = 18.57 \text{ tm} = M_{3,\max}$$

The redundants X_n, span moments M_n, and shear V_n computed by the use of Table 1, $w_D = w_L = 1$ at all spans ($c = 0.25$), are presented in Fig. 2-28. For comparison, see also the moments computed by the zero-point method ($w = 1$ t/m), given in Fig. 2-4.

Concentrated load: $P = 1.0$ t

Span 1:

$$L_1 = 10.00 \text{ m} \qquad x = 3.00 \text{ m}$$

Critical Live Load Locations

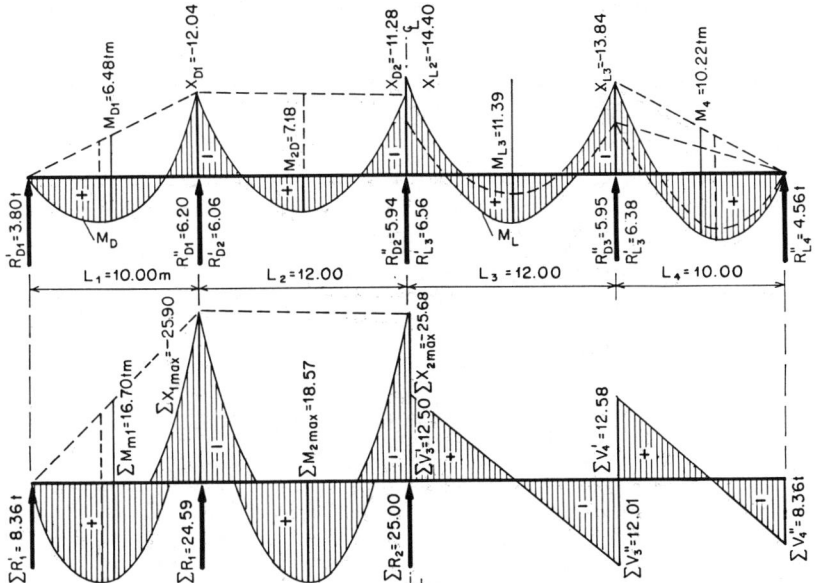

Figure 2-28

$$\xi = \frac{x}{L_1} = \frac{3.00}{10.00} = 0.30 \qquad \xi' = 1.0 - \xi = 0.70$$

$$X_1 = -0.0682\,PL_1 = -0.682 \text{ tm} \qquad R'_1 = 0.632\,P = 0.632 \text{ t}$$

$$M_{1\xi} = 0.1895\,PL_1 = 1.895 \text{ tm} \qquad R''_1 = 0.368\,P = 0.368 \text{ t}$$

Intermediate span n:

$$L_n = 12.00 \text{ m} \qquad x = 3.60 \text{ m}$$

$$\xi_n = \frac{3.60}{12.00} = 0.30 \qquad \xi'_n = 0.70$$

$$X'_n = -0.0770 \cdot PL_n = -0.92 \text{ tm} \qquad R'_n = 0.728\,P = 0.728 \text{ t}$$

$$X''_n = -0.0490 \cdot PL_n = -0.59 \text{ tm} \qquad R''_n = 0.272\,P = 0.272 \text{ t}$$

$$M_{n\xi} = 0.1414 \cdot PL_n = 1.70 \text{ tm}$$

For several loads P at spans, the final redundants X_n and moments M_n are obtained by superposition.

MOMENTS AND REACTIONS IN SHORT SPAN SLABS

For relatively short-span slabs supported and poured simultaneously with beams, the moments and reactions are adequately accurate, as shown in Fig. 2-29.

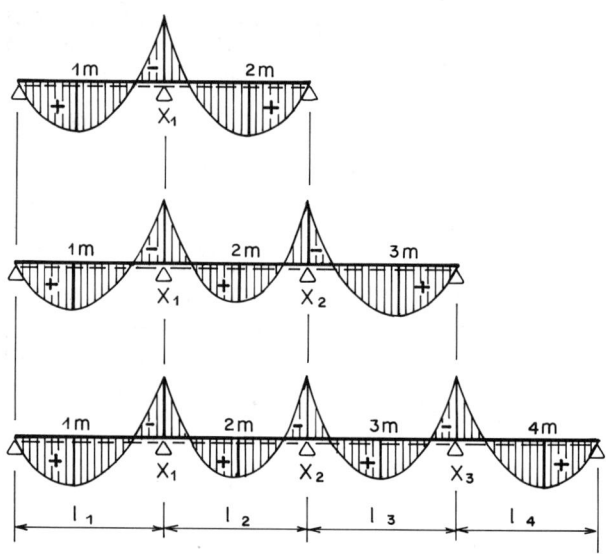

Figure 2-29

Moments:

Two spans: $X_1 = -\dfrac{1}{8} wL_1^2$, $\quad M_{m1} = \dfrac{1}{11} wL_1^2$, $\quad M_{2m} = \dfrac{1}{11} wL_2^2$

Three spans: $X_1 = X_2 = -\dfrac{1}{9} wL_1^2 \quad M_1 = \dfrac{1}{11} wL_1^2 \quad M_3 = \dfrac{1}{11} wL_3^2$

$$M_{m2} = \dfrac{1}{15} wL_2^2 \tag{2-26}$$

Multispans: $X_1 = X_3 = -\dfrac{1}{9} wL^2 \quad M_{m1} = \dfrac{1}{11} wL_1^2 \quad M_{m4} = \dfrac{1}{11} wL_4^2$

$\quad X_2 = -\dfrac{1}{10} wL_2^2 \quad M_{m2} = \dfrac{1}{15} wL_2^2 = M_{m3}$

Reactions:

End spans: $R_1' = 0.40\ wL_1 \quad R_1'' = 0.60\ wL_1 = V_1''$

$\qquad R_n' = 0.60\ wL_n = V_n' \quad R_n'' = 0.40\ wL_n = V_n'$

Intermediate $R'_n = R''_n = 0.50\ wL_n$ (2-27)

spans: $R_1 = (0.60\ L_1 + 0.50\ L_2)w$

$R_n = (0.50 + 0.50)\ wL_n$

3

One- and Multistory Frames

GENERAL

Frames consist of horizontal beams rigidly or elastically connected to columns. The columns are commonly fixed or hinged at the foundation.

The structural behavior of frames is characterized by the fact that the angle between the framing members remains unchanged after deformation of the members—that is, the rotations of the ends (φ) of joining members are equal and also the sum of bending moments at any point is zero (see Fig. 1-3).

There are numerous types of frames, as shown in Fig. 3-1.

Zero-Point Method

All frames are many times statically indeterminate. In order to be able to determine the moments (M), thrust (H), normal force (N), and reactions (R) for dimensioning, the zero-point method is commonly used.

The rigorous frame theories and test results indicate that the end moments of columns M_u'' and M_S' do not appreciably influence the beam moments. Thus, for vertical loading only one floor between the two adjacent floors may be considered for statical analysis.

General

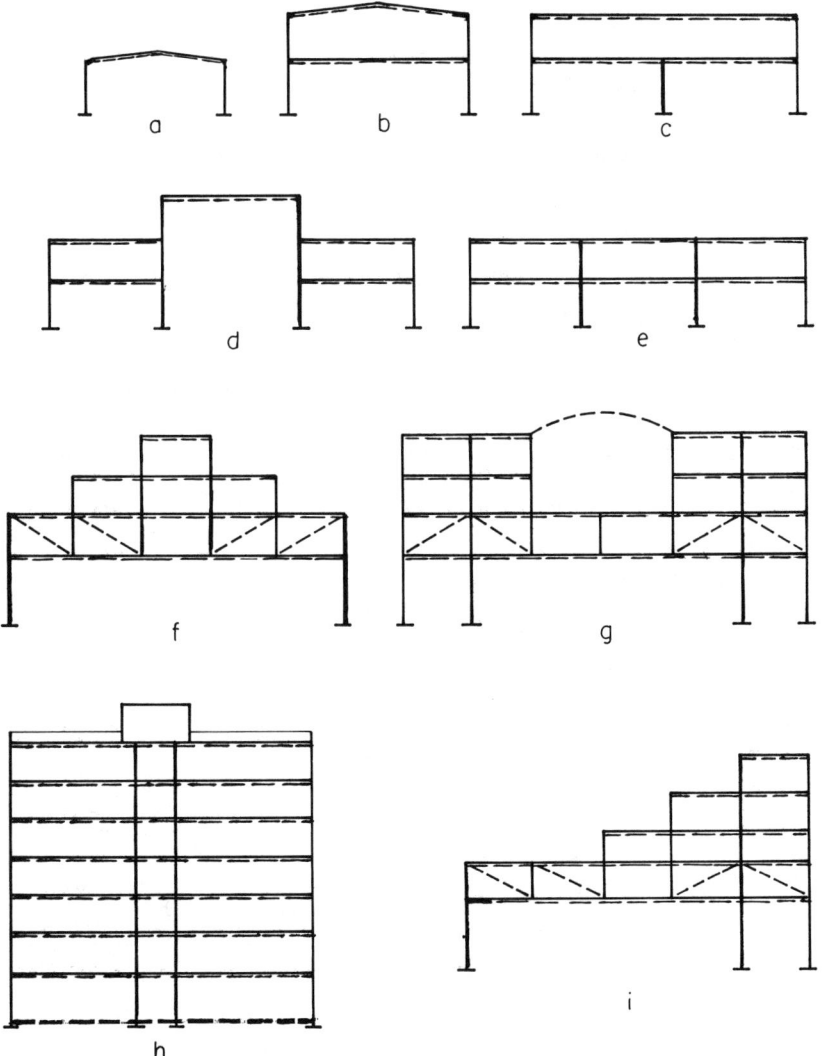

Figure 3-1

By the use of zero-point method, the statically indeterminate frame is converted into statically determinate elements acting between the zero points. The location of zero points is estimated or assumed.

To obtain a structural feeling, the deformations and forces of such elements subjected to loading (w.P.) will be discussed in the following paragraphs.

The one-story frames, shown in Fig. 3-2, are hinged at the base. They are not stable unless laterally supported at the top.

For example,

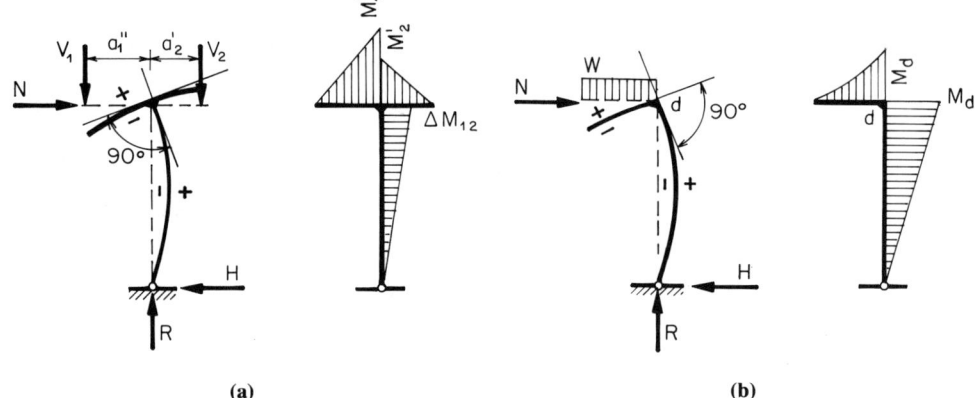

Figure 3-2

Frame type *a*:

$$a_1'' = 3.00 \text{ m}, \, a_2' = 2.00 \text{ m}, \, h = 6.00 \text{ m}$$

$$V_1 = 3.00 \text{ t}, \, V_2 = 2.00 \text{ t}$$

$$M_1'' = a_1'' V_1 = 3.00 \cdot 3.00 = 9.00 \text{ tm}$$

$$M_2' = a_2' V_2 = 2.00 \cdot 2.00 = 4.00 \text{ tm}$$

$$\Delta M = M_1'' - M_2' = 5.00 \text{ tm}$$

$$H = \frac{\Delta M}{h} = \frac{5.00}{6.00} = 0.833 \text{ t}$$

$$\Sigma X = 0: \quad -H = N$$

$$\Sigma Z = 0: \quad V_1 + V_2 = R = 5.00 \text{ t}$$

Frame type *b*:

$$M_d = \frac{w \cdot a_1''^2}{2} = \frac{1.0 \cdot 3.00^2}{2} = 4.50 \text{ tm}$$

$$H = \frac{M_d}{h} = \frac{4.50}{6.00} = 0.75 \text{ t} = N$$

$$R = w \cdot a_1'' = 1.0 \cdot 3.00 = 3.00 \text{ t}$$

The frames illustrated in Figs. 3-3a and b are fixed at the foundation and are stable without lateral support; therefore, $\Delta x \neq 0$, $H = N = 0$. Using the same *a*- and

General

V-values as for hinged frames (Fig. 3-2), the moments and reactions are the same. However, the moments along verticals are constant. The deformations and moments are illustrated in Fig. 3-3.

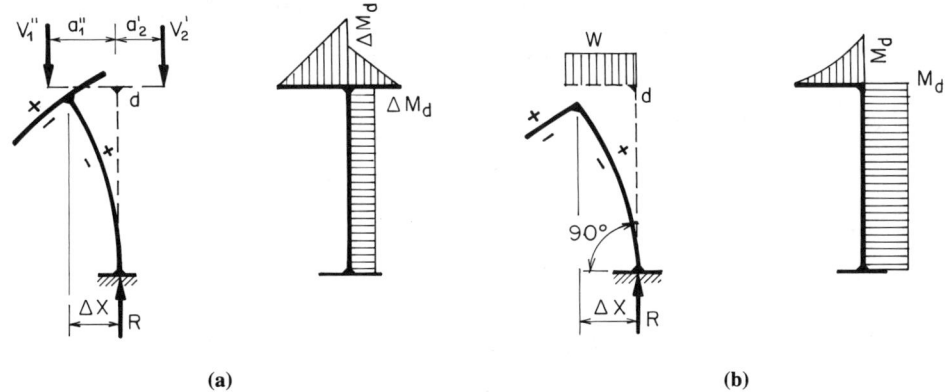

Figure 3-3

The lateral deflection Δx can be determined by Eq. 1-16.

The frames shown in Figs. 3-4a and 3-4b are fixed at the foundation and laterally restrained—they are statically indeterminate. The inflection point (i.p.) of the deflection curvature of the verticals is assumed to be about $h_B = \frac{1}{3} h$ and $h_T = \frac{2}{3} h$. The frame above the i.p. acts as a hinged frame (Fig. 3-2) supported by a cantilever from the foundation.

Thus, the thrust H and base moments M_B will now be computed.

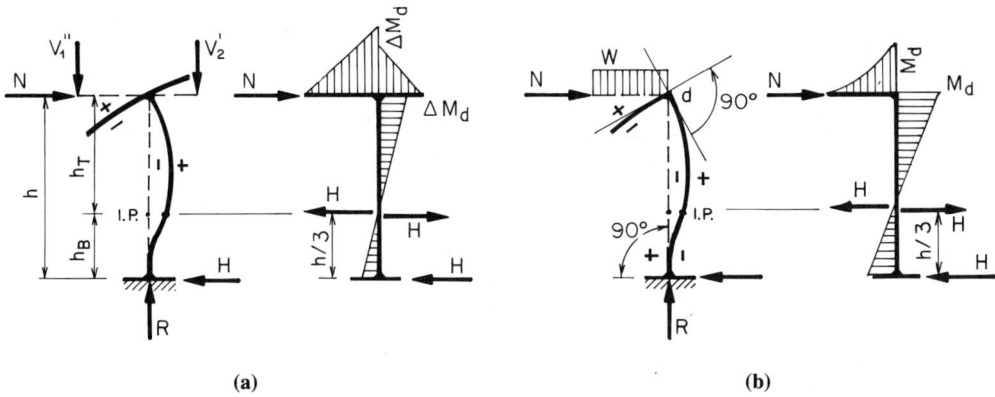

Figure 3-4

Frame type a:

$$H = \frac{\Delta M}{h_T} = \frac{5.00}{0.66 \cdot 6.00} \cong 1.25 \text{ t} = N$$

$$M_B = Hh_B = 1.25 \cdot 0.34 \cdot 6.00 = 2.50 \text{ tm}$$

$$R = V_1'' + V_2' = 3.00 + 2.00 = 5.00 \text{ t}$$

Frame type b:

$$M_d'' = \frac{1.0 \cdot 3.00^2}{2} = 4.50 \text{ tm}$$

$$H = \frac{M_d''}{h_T} = \frac{4.50}{4.00} = 1.125 \text{ t}$$

$$M_B = Hh_B = 1.125 \cdot 2.00 = 2.25 \text{ tm}$$

$$R = wa_2'' = 1.0 \cdot 3.00 = 3.00 \text{ t}$$

The moments are illustrated in Fig. 3-4.

The frames or part of frames, illustrated in Fig. 3-5, demonstrate the frame-carrying action and deformation under uniform load w.

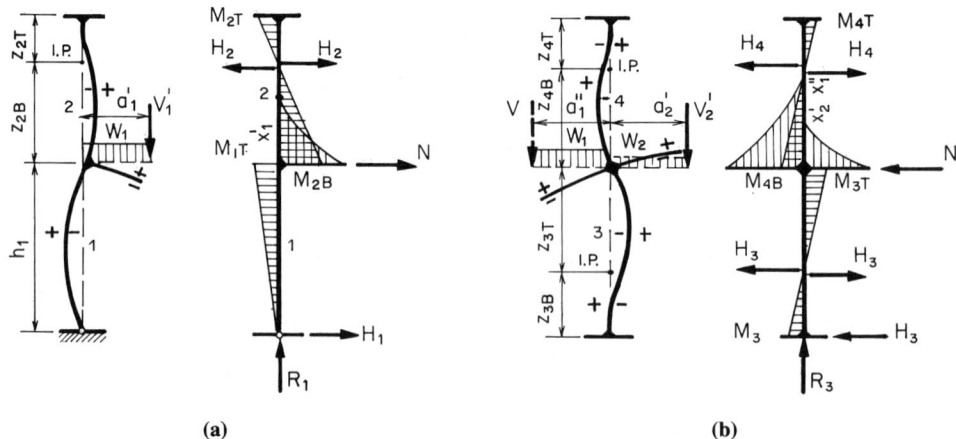

(a)   (b)

Figure 3-5

Frame type a:

Cantilever: $L_1 = 10.00$ m $a_1' = 0.10 L_1 = 1.00$ m

$a_1'' = 0.23 L_1 = 2.30$ m

General

$$L_{10} = L_1 - (a'_1 + a''_1) = 10.00 - (1.00 + 2.30) = 6.70 \text{ m}$$

$$w_1 = 1.0 \text{ t/m}$$

$$V'_1 = \frac{w_1 \cdot L_{10}}{2} = \frac{1.0 \cdot 6.70}{2} = 3.35 \text{ t}$$

$$X'_1 = V'_1 a'_1 + \frac{w_1 a'^2_1}{2} = 3.35 \cdot 1.00 + \frac{1.0 \cdot 1.00^2}{2} = 3.85 \text{ tm}$$

Columns: $h_1 = 4.00$ m, $z_{1B} = 0$

$h_2 = 3.30$ m, $z_{2T} = 0.285 \cdot h_2 = 0.94$ m

$$z_{2B} = 3.30 - 0.94 = 2.36 \text{ m}$$

$I_1 = 1.0$, $I_2 = 0.8\, I_1 = 0.80$

$$S_1 = \frac{I_1}{h_1} = \frac{1.0}{4.00} = 0.25$$

$$S_2 = \frac{I_2}{z_{2B}} = \frac{0.80}{2.36} = 0.34$$

$$M_{1T} = X'_1 \frac{S_1}{S_1 + S_2} = 3.85 \frac{0.25}{0.25 + 0.34} = 1.63 \text{ tm}$$

$$M_{2B} = X'_1 \frac{S_2}{S_1 + S_2} = 3.85 \frac{0.34}{0.25 + 0.34} = 2.22 \text{ tm}$$

$M_{1B} = 0$

$M_{2T} = c_T M_{2B} = 0.40 \cdot 2.22 = 0.89$ tm

$$H_1 = \frac{M_{1T}}{h_1} = \frac{1.63}{4.00} = 0.41 \text{ t}$$

$$H_2 = \frac{M_{2B}}{z_{2B}} = \frac{2.22}{2.36} = -0.94 \text{ t}$$

$\Sigma X = 0$: $N = H_2 - H_1 = 0.94 - 0.41 = 0.53$ t

$\Sigma Z = 0$: $R_1 = V'_1 + a'_1 w_1 = 3.35 + 1.00 \cdot 1.0 = 4.35$ t

Frame type *b*:

Cantilevers: $L_2 = 10.00$ m $\qquad a'_2 = a''_2 = 0.20\, L_2 = 2.00$ m

$L_{o2} = L_2 - (a'_2 + a''_2) = 6.00$ m

$w_2 = 0.7$ t/m

$$V_2' = \frac{w_2 L_{20}}{2} = \frac{0.7 \cdot 6.00}{2} = 2.10 \text{ t}$$

$$X_2' = V_2' a_2' + \frac{w_2 a_1'^2}{2} = 2.10 \cdot 2.00 + \frac{0.70 \cdot 2.00^2}{2} = 5.60 \text{ tm}$$

$$X_1'' = V_1'' a_1'' + \frac{w_1 a_1''^2}{2} = 3.35 \cdot 2.30 + \frac{1.0 \cdot 2.30^2}{2} = 10.35 \text{ tm}$$

$$\Delta X_1 = X_1'' - X_2' = 10.35 - 5.60 = 4.75 \text{ tm}$$

Columns: $\quad h_3 = 4.00 \text{ m} \quad z_{3B} = 0.333 \cdot 4.00 = 1.33 \text{ m} \quad z_{3T} = 2.67 \text{ m}$

$\quad\quad\quad\quad h_4 = 3.30 \text{ m} \quad z_{4T} = 0.94 \text{ m}, \quad z_{4B} = 2.36 \text{ m}$

$$I_3 = 2I_1 = 2.0 \quad I_4 = 0.8\, I_3 = 1.6$$

$$S_3 = \frac{I_3}{z_{3T}} = \frac{2.0}{2.67} \cong 0.75$$

$$S_4 = \frac{I_4}{z_{4B}} = \frac{1.6}{2.36} = 0.68$$

$$M_{3T} = \Delta X_1 \frac{S_3}{S_3 + S_4} = 4.75 \frac{0.75}{0.75 + 0.68} = 2.50 \text{ tm}$$

$$M_{4B} = \Delta X_1 \frac{S_4}{S_3 + S_4} = 4.75 \frac{0.68}{0.75 + 0.68} = 2.25 \text{ tm}$$

$$M_{3B} = c_3 M_{3T} = 0.5 \cdot 2.50 = 1.25 \text{ tm},$$

$$M_{4T} = c_4 M_{4B} = 0.4 \cdot 2.25 = 0.90 \text{ tm}$$

$$H_3 = \frac{M_{3T}}{z_{3T}} \cong 0.94 \text{ t} \quad\quad H_4 = \frac{M_{4B}}{z_{4B}} = 0.96 \text{ t}$$

$$N = H_4 - H_3 = 0.02 \text{ t}$$

$$R_3 = V_1'' + w_1 a_1'' + V_2' + w_2 a_2' = 9.15 \text{ t}$$

Loading Factor Method

As can be seen from discussed examples, the zero point method is rather simple. The difficulties are to estimate the zero point distances (a) accurately because they depend on the stiffnesses (S) of the joining members. The most practical method to obtain qualitatively and quantitatively acceptable results is the loading factor method, similar to the one used to determine the moments of the continuous beams. The loading factor method for frames will be discussed in the following sections.

MULTISTORY FRAMES

Uniform Vertical Loading

For simplicity, span-by-span loading for frames and vertical loading $w = 1$ t/m for spans is used. The deformations, moments (M, X), and shear (V) of a three-span frame loaded at the center span (L_2) are shown in Fig. 3-6.

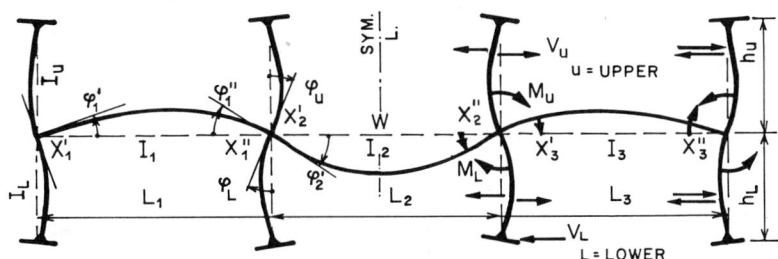

Figure 3-6

Denoting the distribution factor with d and carryover by c and considering that the loading factors $\overline{L}$ for unloaded spans and columns are zero, in accordance with Eqs. 2-14 and 2-15 the end rotations

$$\varphi_{n-1} = \varphi_n = \varphi_L = \varphi_u$$

and their sum $\Sigma \varphi = 0$.

The stiffness factors (S) for spans (S_n) and columns (S_L, S_U) are:

$$S_n = \frac{I_n}{L_n} \qquad S_L = \frac{I_L}{h_L} \text{ (lower columns)}$$

$$S_u = \frac{I_u}{h_u} \text{ (upper columns)} \tag{3-1}$$

The zero-point locations (a) of columns, stated before, do not appreciably influence the beam moments and, therefore, may be taken for elastically restrained ends $a'_L = 0.285\ h_L$ and $a''_u = 0.285\ h_u$, thus in accordance with Eq. 2-21 the carryovers (c) for columns $c'_L = c''_u = 0.4$.

Where column ends at foundations are rigidly constrained, a'_L–may be taken as $0.33\ h_L$. Thus, $c'_L = 0.5$.

Distribution factors.

$$6\varphi''_{n-1} = \frac{L_{n-1}}{I_{n-1}}(2X'' + X') \qquad \overline{L}_1 = 0 \qquad \text{(Eq. 2-14)}$$

$$-X'_{n-1} = c'_{n-1} X''_{n-1} \qquad \frac{L_{n-1}}{I_{n-1}} = S_{n-1}$$

$$6\varphi'' = S_{n-1}(2 - c'_{n-1}) X''_{n-1} = \frac{X''_{n-1}}{d''_{n-1}}$$

Thus, the distribution factors:

$$d''_{n-1} = \frac{S_{n-1}}{2 - c'_{n-1}} \quad - \text{ beams} \tag{3-2}$$

Similarly

$$d'_u = \frac{S_u}{2 - c_u} \qquad S_u = \frac{h_u}{I_u} \quad - \text{ upper columns}$$

$$d''_L = \frac{S_L}{2 - c_L} \qquad S_L = \frac{h_L}{I_L} \quad - \text{ lower columns}$$

The sum of the distribution factors for each of the beam ends is:

$$D''_{n-1} = d''_{n-1} + d'_u + d''_L$$
$$D'_n = d'_n + d'_u + d''_L \tag{3-3}$$

Carryover factors. Equilibrium requires that the sum of moments (M, X) at any joint must be zero; thus (Fig. 3-6),

$$X''_{n-1} + X''_L + X'_u = X'_n \tag{3-4}$$

In accordance with Eqs. 3-3 and 3-4 and considering

$$\frac{X''_{n-1}}{X''_L} = \frac{d''_{n-1}}{d''_L} \qquad \frac{X''_{n-1}}{X'_u} = \frac{d''_{n-1}}{d'_u}$$

we obtain:

$$\frac{X''_{n-1}}{X''_{n-1} + X''_L + X'_u} = \frac{X''_{n-1}}{X'_n} = \frac{d''_{n-1}}{d''_{n-1} + d''_L + d'_u} = \frac{d''_{n-1}}{D''_{n-1}}$$

$$X''_{n-1} = \frac{d''_{n-1}}{D''_{n-1}} X'_n$$

$$X'_{n-1} = c'_{n-1} X''_{n-1} \tag{3-5}$$

$$X'_{n+1} = \frac{d'_{n+1}}{D'_{n+1}} X''_n$$

$$X''_{n+1} = c''_{n+1} X'_{n+1}$$

Multistory Frames

Considering that $\varphi'_n = -\varphi''_{n-1}$, then by Eqs. 2-14 and 3-5 we obtain the carryovers (c)

From 0 to 4: $\qquad c'_n = \dfrac{D''_{n-1}}{2D''_{n-1} + S_n}$

From 4 to 0: $\qquad c''_n = \dfrac{D'_n}{2D'_n + S_{n-1}}$

(3-6)

The carryover (c), distribution (d), and $D = \Sigma d$ are presented in Fig. 3-7a.

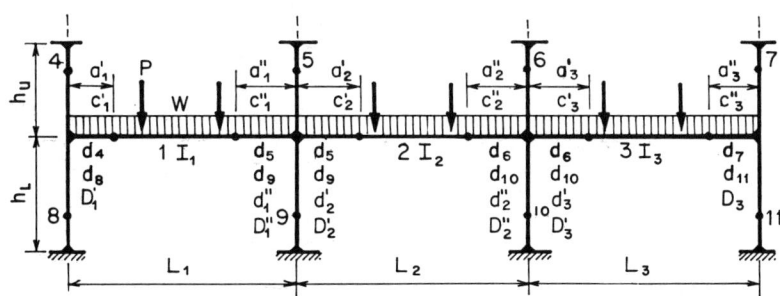

Figure 3-7(a)

The beam end moments (X', X'') are computed by the same equations developed for continuous beams (Eqs. 2-19 and 2-20), except the carryover factors (c', c'') differ. Thus,

$\overline{L}'_n \neq \overline{L}''_n$: $\qquad X'_n = -c'_n \dfrac{\overline{L}'_n - c''_n \overline{L}''_n}{1 - c'_n c''_n}$

(Eq. 2-19)

$\qquad\qquad\qquad X''_n = -c''_n \dfrac{\overline{L}''_n - c'_n \overline{L}'_n}{1 - c'_n c''_n}$

$\overline{L}'_n = \overline{L}''_n$: $\qquad X'_n = -c'_n \overline{L}'_n \dfrac{1 - c''_n \overline{L}''_n}{1 - c'_n c''_n}$

(Eq. 2-20)

$\qquad\qquad\qquad X''_n = -c''_n \overline{L}''_n \dfrac{1 - c'_n \overline{L}'_n}{1 - c'_n c''_n}$

Column Moments

For simplicity, in practice the column end moments (X''_L, X'_u) are computed from the difference of the beam end moments (Eq. 3-5).
Outside columns:

$$X''_L = X'_1 \dfrac{d_L}{d_L + d_u} \qquad X'_L = X''_L c'_L$$

(3-7)

$$X'_u = X'_1 \frac{d_u}{d_L + d_u} \qquad X''_u = X'_u c''_u$$

Inside columns:

$$\Delta X_n = \Sigma X'_n - \Sigma X''_{n-1}$$

$$X''_L = \Delta X_n \frac{d_L}{d_L + d_u} \qquad X'_L = X''_L c'_L \qquad (3\text{-}8)$$

$$X'_u = \Delta X_n \frac{d_u}{d_L + d_u} \qquad X''_u = X'_u c''_u$$

The final beam and column end moments (X_n, X_L, X_u), thrusts (H), normal forces (N), and reactions are:

$$\Sigma X'_n, \ \Sigma X''_{n-1}, \ \Sigma X'_n$$

$$\Sigma X''_{nL}, \ \Sigma X'_{nL}, \ \Sigma X'_{nu}, \ \Sigma X''_{nu}$$

$$\Sigma H_{nL} = \Sigma \frac{X''_L}{a''_L}, \quad \Sigma H_{nu} = \Sigma \frac{X'_u}{a'_u} \qquad (3\text{-}9)$$

$$\Sigma N_n = \Sigma H_{nL} - \Sigma H_{nu}$$

$$\Sigma R_0 = \Sigma R'_1$$

$$\Sigma R_{n-1} = \Sigma R''_{n-1} + \Sigma R'_n$$

Example

For illustration of the loading factor method for frames (Fig. 3-7b) subjected to vertical loads, $(w, 2P)$ will be computed.

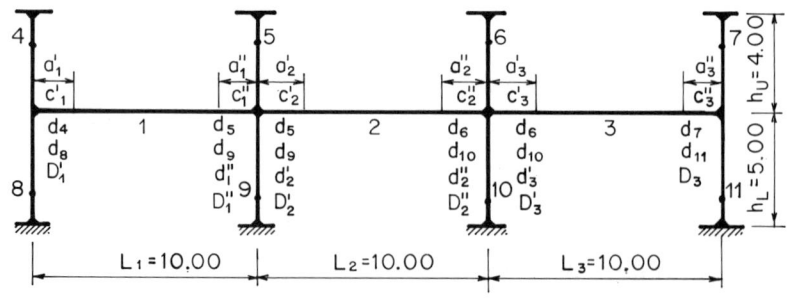

Figure 3-7(b)

By expressing the I–values of members in ratios I_h/I_n, the inaccuracies are reduced in large degree.

$$I_1 = I_2 = I_n = 10 \text{ dm}^4 \qquad \text{dm} = 10 \text{ cm}$$

$$I_{4,7} = 1.55 \text{ dm}^4 \qquad I_{5,6} = 3.65 \text{ dm}^4$$

Multistory Frames

$$I_{8,11} = 2.44 \text{ dm}^4 \qquad I_{9,10} = 5.45 \text{ dm}^4$$

Stiffnesses:
$$S_n = \frac{I_n}{L_n} = \frac{10}{10} = 1.00$$

$$S_{4,7} = \frac{1.55}{4.00} = 0.39 \qquad S_{5,6} = \frac{3.65}{4.00} = 0.91$$

$$S_{8,11} = \frac{2.44}{5.00} = 0.49 \qquad S_{9,10} = \frac{5.45}{5.00} = 1.09$$

Distribution factors for columns:

$$d_{4,7} = \frac{S_4}{2 - c_4} = \frac{0.39}{2 - 0.4} = 0.244 \qquad d_{5,6} = \frac{0.91}{2 - 0.4} = 0.568$$

$$d_{8,11} = \frac{S_8}{2 - c_5} = \frac{0.49}{2 - 0.5} = 0.327 \qquad d_{9,10} = \frac{1.09}{2 - 0.5} = 0.727$$

Carryovers (c'_n), distribution factors (d'_n), and D'_n-values are obtained by starting from the left and $c''_n d''_n$ and D''_n starting from the right. They are:

$$c'_1 = \frac{1}{2 + \frac{S_1}{S_4 + S_8}} = \frac{1}{2 + \frac{1.000}{0.244 + 0.327}} = 0.266 = c''_3 \qquad D'_1 = d_{4,7} + d_{8,1}$$
$$= 0.244 + 0.327 = 0.571 = D''_3$$

$$d'_1 = \frac{{}_1 S}{2 - c'_1} = \frac{1.000}{2 - 0.266} = 0.577 = d''_3 \qquad D''_1 = d''_1 + d_5 + d_9 = D'_3$$
$$= 0.577 + 0.568 + 0.727 = 1.872$$

$$c'_2 = \frac{D''_1}{2D''_1 + S_2} = \frac{1.872}{2 \cdot 1.872 + 1.000} = 0.395 = c''_2$$

$$d''_2 = \frac{S_2}{2 - c'_2} = \frac{1,000}{2 - 0.395} = 0.623 = d'_2 \qquad D''_2 = d''_2 + d_6 + d_{10} = D'_2$$
$$= 0.623 + 0.568 + 0.727 = 1.918$$

$$c'_3 = \frac{D''_2}{2 D''_2 + S_3} = \frac{1.918}{2 \cdot 1.918 + 1.000} = 0.397 = c''_1$$

Uniform Loading

$w_1 = w_2 = w_3 = 1.0$ t/m.

Loading factors: $\bar{L}_1 = \bar{L}_2 = \bar{L}_3 = \dfrac{w \cdot L^2}{4} = \dfrac{1.0 \cdot 10.00^2}{4} = 25$ tm

Span 3 and span 1: $w_3 = 1$:

$$X''_3 = X'_1 = -c'_1 \bar{L} \frac{1 - c''_1}{1 - c'_1 c''_1} = -0.266 \cdot 25 \frac{1 - 0.397}{1 - 0.266 \cdot 0.397} = -4.48 \text{ tm}$$

$$X'_3 = X''_1 = -c''_1 \bar{L} \frac{1 - c'_1}{1 - c'_1 c''_1} = -0.397 \cdot 25 \frac{1 - 0.266}{1 - 0.266 \cdot 0.397} = -8.15 \text{ tm}$$

$$X'_2 = \frac{d'_2}{D'_2} X''_1 = -\frac{0.623}{1.918} 8.15 = -2.65 \text{ tm}, \qquad X''_2 = 1.04 \text{ tm}$$

$$X'_3 = \frac{d'_3}{D'_3} X''_2 = \frac{0.577}{1.872} 1.04 = 0.32 \text{ tm}$$

$$X''_3 = c''_3 X'_3 = 0.266 \cdot 0.32 = -0.09 \text{ tm}$$

Span 2: $w_2 = 1$:

$$X'_2 = -c'_2 \bar{L} \frac{1 - c''_2}{1 - c'_2 c''_2} = -0.395 \cdot 25 \frac{1 - 0.395}{1 - 0.395^2} = -7.60 \text{ tm}$$

$$X''_2 = -c''_2 \bar{L} \frac{1 - c'_2}{1 - c'_2 c''_2} = X'_2 = -7.60 \text{ tm}$$

$$X''_1 = -\frac{d''_1}{D_1} X'_2 = \frac{0.577}{1.872} 7.60 = -2.34 \text{ tm} = X'_3$$

$$X'_1 = c'_1 X''_1 = 0.266 \cdot 2.34 = 0.62 \text{ tm}$$

The moments are shown in Figs 3-8 and 3-9.

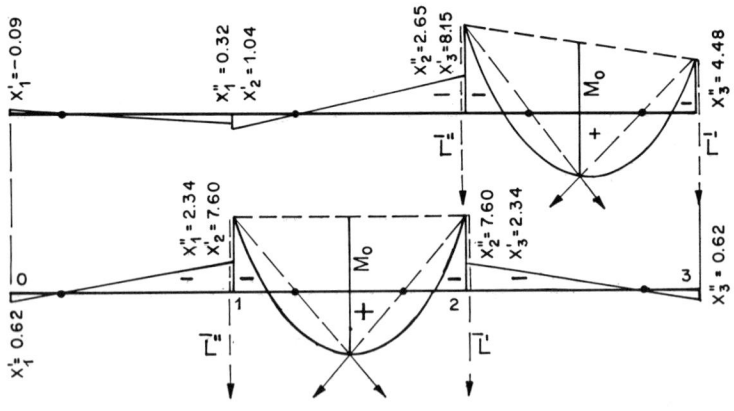

Figures 3-8 and 3-9

Superposition:
Beam moments:

$$\Sigma X'_1 = -(4.48 + 0.09 - 0.62) = 3.95 \text{ tm} = \Sigma X''_3$$

$$\Sigma X''_1 = -(8.15 + 2.34 - 0.32) = -10.17 \text{ tm} = \Sigma X'_3$$

$$\Sigma X'_2 = -(2.64 + 7.60 - 1.04) = -9.20 = \Sigma X''_2$$

$$\Delta X_{1,2} = \Sigma X''_1 - \Sigma X'_2 = -10.17 + 9.20 = -0.97 \text{ tm} = \Delta X_{3,2}$$

Multistory Frames

Column moments:

$$X_8'' = \Sigma X_1' \frac{d_8}{d_8 + d_4} = 3.95 \cdot \frac{0.327}{0.327 + 0.244} \cong -2.27 \text{ tm} = X_{11}''$$

$$X_8' = c_8' X_8'' = 0.5 \cdot 2.27 = 1.14 \text{ tm} = X_{11}'$$

$$X_4' = \Sigma X_1' \frac{d_4}{d_8 + d_4} = 1.69 \text{ tm} = X_7''$$

$$X_4'' = c_4'' X_4' = 0.4 \cdot 1.69 = 0.68 \text{ tm} = X_7''$$

$$X_9'' = \Delta X_2 \frac{d_9}{d_9 + d_5} = 0.97 \cdot \frac{0.727}{0.727 + 0.568} = 0.54 \text{ tm}$$

$$X_9' = c_9' X_9'' = 0.5 \cdot 0.54 = 0.27 \text{ tm} = X_{10}'$$

$$X_5' = \Delta X_2 \frac{d_5}{d_9 + d_5} = 0.97 \cdot \frac{0.568}{0.727 + 0.568} = 0.43 \text{ tm}$$

$$X_5'' = c_5'' X_5' = 0.4 \cdot 0.43 = 0.17 \text{ tm} = X_6''$$

Span moments:

$$M_0 = \frac{1}{8} wL^2 = \frac{1}{8} 1.0 \cdot 10.00^2 = 12.50 \text{ tm}$$

$$M_{1m} = M_0 - \frac{\Sigma X_1' + \Sigma X_1''}{2} = 12.5 - \frac{3.95 + 10.17}{2} = 5.45 \text{ tm}$$

$$M_{2m} = M_0 - \Sigma X_2' = 12.50 - 9.20 = 3.30 \text{ tm}$$

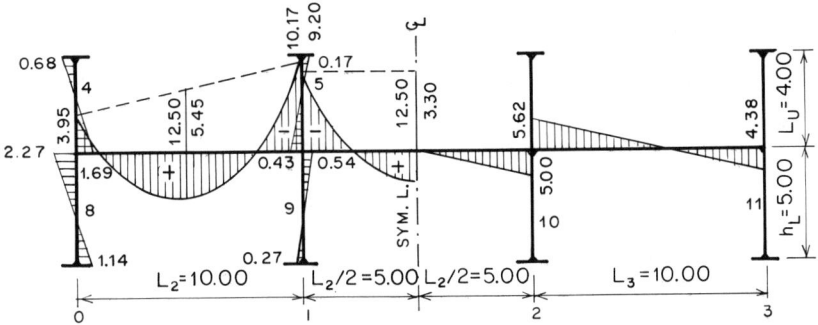

Figure 3-10

Thrusts:

$$a_4' = a_5' = (1 - 0.285)4.00 = 2.86 \text{ m}$$

$$a_8'' = a_9'' = (1 - 0.333)5.00 = 3.33 \text{ m}$$

$$H_4 = \frac{\Sigma X_4'}{a_4'} = -\frac{1.69}{2.86} = -0.590 \text{ t}$$

$$H_8 = \frac{\Sigma X_8''}{a_8''} = \frac{2.27}{3.33} = 0.680 \text{ t}$$

Normal forces:

$$N_1 = H_8 - H_4 = 0.680 - 0.590 = 0.090 \text{ t}$$
$$= N_3$$

Reactions:

$$R_1' = R_0 - \frac{\Sigma X_1' + \Sigma X_1''}{L_1} = 5.00 - \frac{10.17 - 3.95}{10.00}$$

$$= 4.38 \text{ t} = R_3'' \qquad R_1'' = 5.00 + 0.62 = 5.62 \text{ t}$$

$$R_1 = R_1'' + R_2' = 5.62 + 5.0 = 10.62 \text{ t} = R_2$$

Concentrated Loads

The moments and reactions for concentrated loads ($2P$ at spans) can be obtained by multiplying the moments and reactions due to uniform load (w) (Fig. 3-10) by the factor (r).

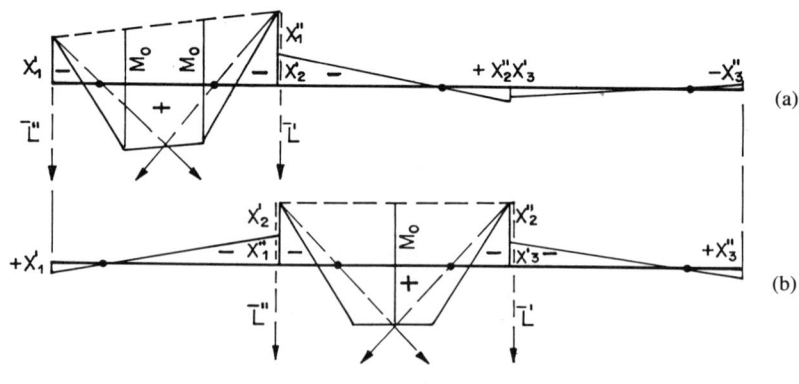

Figure 3-11

$$\overline{L}_W = \frac{wL^2}{4} = 25.00 \text{ tm}$$

$$\overline{L}_{2P} = 0.666 \cdot L = 6.66 \text{ tm}, \quad r = \frac{\overline{L}_{2P}}{\overline{L}_W} = \frac{6.66}{25.00} = 0.267$$

Multistory Frames

The span moments are illustrated in Fig. 3-11. To obtain maximum moments and reactions, the live-load (w_L, P_L) locations are given in Fig. 2-27.

Horizontal Loading

The zero points in columns are assumed to vary linearly from $a_1 = 0.55\,h_1$ to $a_n = 0.33\,h_n$ (lowest story). However, for simplicity they are commonly taken a_1 to $a_{n-1} = 0.5\,h$.

The zero points in span, because of symmetry, are $a' = a'' = L/2$

$$\sum_1^n W = W_1 + W_2 + W_3 \cdots + W_n$$

$$M_n = W_1(h_1 + h_2 \cdots + h_{n-1} + a_n) + W_2(h_2 + h_3 \cdots \qquad (3\text{-}10)$$
$$+ h_{n-1} + a_n) + W_3(h_3 + h_4 \cdots h_{n-1} + a_n) + W_n a_n$$

The increase of the moment ΔM between two zero points is

$$\Delta M_{n,n-1} = M'_n - M'_{n-1}$$
$$\Delta M_{n+1,n} = M'_{n+1} - M'_n \qquad (3\text{-}11)$$

The shear and moment diagrams are plotted in Fig. 3-12a.

The forces and moments are computed by the three equations of equilibrium $\Sigma Z = 0 \quad \Sigma X = 0 \quad$ and $\Sigma M = 0$.

One-bay frame

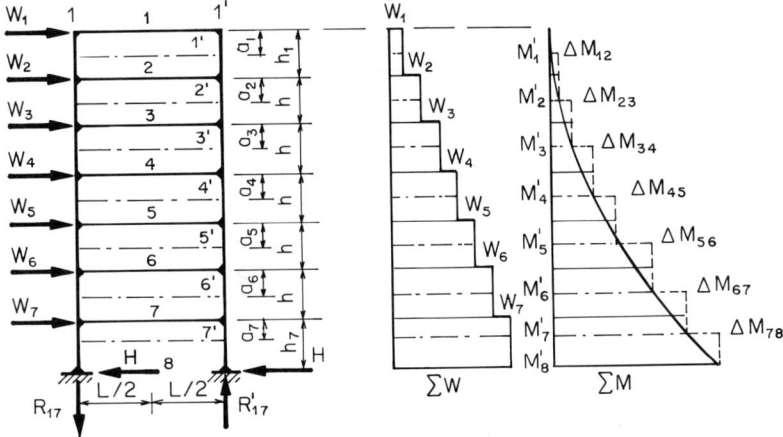

Figure 3-12(a)

$$R_{1n} - V_1 - V_2 \cdots - V_n = 0$$

$$H_{1n} + N_1 + N_2 \cdots + N_n - \sum_1^n W = 0$$

$$M'_n - R_{1n}L = 0$$ (3-12)

$$2H_n - \sum_1^n W = 0$$

The forces and moments acting upon the one-bay frame are illustrated in Fig. 3-12b.

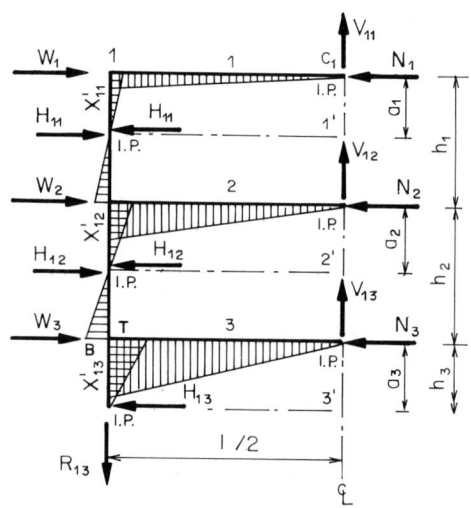

Figure 3-12(b)

$$X'_{11} = \frac{1}{2} M'_1 \qquad V_{11} = \frac{M'_1}{L}$$

$$X_{12} = \frac{1}{2} \Delta M_{12} \qquad V_{12} = \frac{\Delta M_{12}}{L}$$

$$X_{13} = \frac{1}{2} \Delta M_{13} \qquad V_{13} = \frac{\Delta M_{13}}{L}$$

$$H_{11} = \frac{1}{2} W_1 \qquad M_{11} = H_{11}a_1$$

$$H_{12} = \frac{1}{2} \sum_1^2 W \qquad M_{12} = H_{12}a_2 \qquad (3\text{-}13)$$

$$H_{13} = \frac{1}{2} \sum_1^3 W \qquad M_{13} = H_{13}a_3$$

Multistory Frames

$$N_1 = \frac{1}{2}W_1 \qquad R_{11} = \frac{M'_1}{L}$$

$$N_2 = \frac{1}{2}W_2 \qquad R_{12} = \frac{M'_2}{L}$$

$$N_3 = \frac{1}{2}W_3 \qquad R_{13} = \frac{M'_3}{L}$$

Two-bay frames. A two-bay frame subjected to horizontal loads W is shown in Fig. 3-13a. The moments and shear due to W and the increase of moment ΔM between two adjacent-column zero points are computed by Eqs. 3-10 and 3-11 (for the moment-shear diagram, see Fig. 3-12a).

Equilibrium requires:

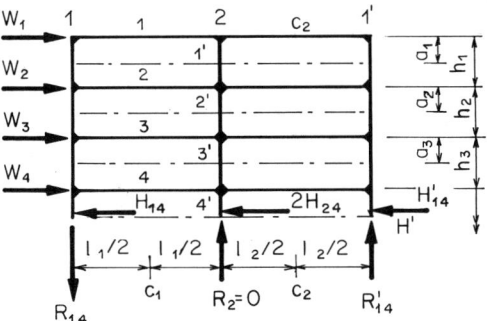

Figure 3-13(a)

$$R_{1n} - V_{1n} - V_{12} \cdots - V_{1n} = 0$$
$$H_{1n} + N_{11} + N_{12} \cdots + N_{1n} - \sum_{1}^{n} W = 0 \qquad (3\text{-}14)$$
$$M'_n - R_{1n}2L = 0$$
$$4H_{1n} - \sum_{1}^{n} W = 0$$

The forces and moments acting upon the two-bay frame are illustrated in Fig. 3-13b.

$$V_{11} = V_{21} = \frac{M'_1}{2L} \qquad X'_{11} = \frac{1}{4}M'_1 = -X''_{11} = X'_{21} = -X''_{21}$$

$$V_{12} = V_{22} = \frac{\Delta M_{12}}{2L} \qquad X'_{12} = \frac{1}{4}\Delta M_{12} = -X''_{12} = X'_{22} = -X''_{22}$$

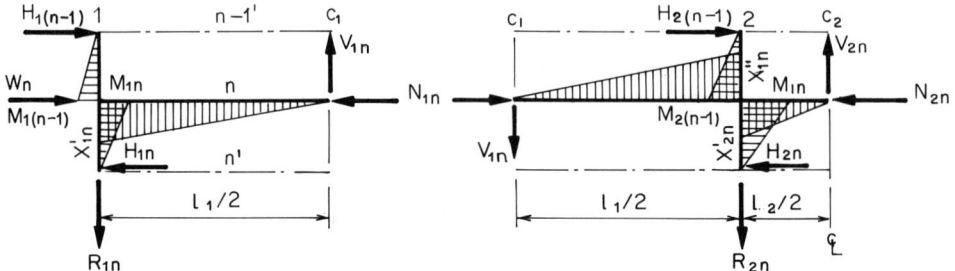

Figure 3-13(b)

$$V_{13} = V_{23} = \frac{\Delta M_{23}}{2L} \qquad X'_{13} = \frac{1}{4}\Delta M_{23} = -X''_{13} = X'_{23} = -X''_{23}$$

$$V_{14} = V_{24} = \frac{\Delta M_{34}}{2L} \qquad X'_{14} = \frac{1}{4}\Delta M_{34} = -X''_{14} = X'_{24} = -X''_{24}$$

$$H_{11} = \frac{1}{4}W_1 = H'_{21} \qquad M_{11} = H_{11}a_1 \qquad H_{21} = 2H_{11} \qquad M_{21} = H_{21}a_1$$

$$H_{12} = \frac{1}{4}\sum_1^2 W = H'_{22} \qquad M_{12} = H_{12}a_2 \qquad H_{22} = 2H_{12} \qquad M_{22} = H_{22}a_2 \qquad (3\text{-}15)$$

$$H_{13} = \frac{1}{4}\sum_1^3 W = H'_{23} \qquad M_{13} = H_{13}a_3 \qquad H_{23} = 2H_{13} \qquad M_{23} = H_{23}a_3$$

$$H_{14} = \frac{1}{4}\sum_1^4 W = H'_{24} \qquad M_{14} = H_{14}a_4 \qquad H_{24} = 2H_{14} \qquad M_{24} = H_{24}a_4$$

$$R_{11} = -\frac{M'_1}{2L} = R'_{21} \qquad\qquad N_{11} = \frac{3}{4}W_1 \qquad N_{21} = \frac{1}{4}W_1$$

$$R_{12} = -\frac{M'_2}{2L} = R'_{22} \qquad R_{2n} = 0 \qquad N_{12} = \frac{3}{4}W_2 \qquad N_{22} = \frac{1}{4}W_2$$

$$R_{13} = -\frac{M'_3}{2L} = R'_{23} \qquad\qquad N_{13} = \frac{3}{4}W_3 \qquad N_{23} = \frac{1}{4}W_3$$

$$R_{14} = -\frac{M'_4}{2L} = R'_{24} \qquad\qquad N_{14} = \frac{3}{4}W_4 \qquad N_{24} = \frac{1}{4}W_4$$

Three-bay frame. In a three-bay frame, shown in Fig. 3-14a, the conditions of equilibrium are:

$$R_{1n}(2L_1 + L_2) + R_{2n}L_2 - M'_n = 0$$

Multistory Frames

$$2(H_{1n} + H_{2n}) - \sum_1^n W = 0$$

$$V_{11} + V_{12} \cdots + V_{1n} - \sum_1^n W = 0$$

$$N_{11} + N_{12} \cdots + N_{1n} + H_{1n} - \sum_1^n W = 0$$

$$N_{21} + N_{22} \cdots + N_{2n} + (H_{1n} + H_{2n}) - \sum_1^n W = 0 \quad (3\text{-}16)$$

$$V_{21} + V_{22} \cdots + V_{2n} - (R_{1n} + R_{2n}) = 0$$

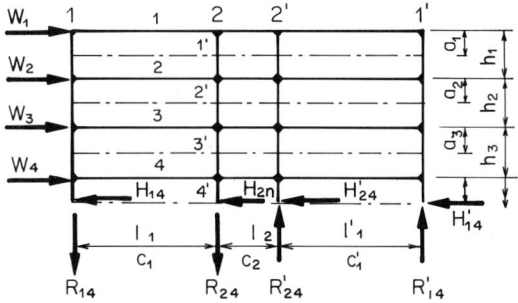

Figure 3-14(a)

According to frame theory, the relationship between L_1 and L_2 is assumed to be

$$R_{1n} = \alpha R_{2n} \quad \alpha \cong \frac{L_1 I_2}{2L_2 I_1} \quad (3\text{-}17)$$

where α is the average moment of inertia of column lines 1 and 2. Thus,

$$R_n = \frac{M_n}{\overline{L}} \qquad \overline{L} = 2L_1 + L_2(1 + \alpha) \quad (3\text{-}18)$$

The forces and moments computed by Eqs. 3-16, 3-17, and 3-18 are plotted and indicated in Fig. 3-14b.

$$V_{11} = \frac{M_1'}{\overline{L}} \qquad V_{21} = (1 + \alpha)V_{11}$$

$$V_{12} = \frac{\Delta M_{12}}{\overline{L}} \qquad V_{22} = (1 + \alpha)V_{12}$$

$$V_{1n} = \frac{\Delta M_{1n}}{\overline{L}} \qquad V_{2n} = (1 + \alpha)V_{1n}$$

$$X'_{11} = -X''_{11} = \frac{L_1}{2} V_{11} \qquad X'_{21} = -X''_{21} = \frac{L_2}{2} V_{21}$$

$$X'_{12} = -X''_{11} = \frac{L_1}{2} V_{12} \qquad X'_{22} = -X''_{22} = \frac{L_2}{2} V_{22}$$

$$X'_{1n} = -X''_{1n} = \frac{L_1}{2} V_{1n} \qquad X'_{2n} = -X''_{2n} = \frac{L_2}{2} V_{2n}$$

$$R_{11} = \frac{M'_1}{L} \qquad R_{21} = \alpha R_{11}$$

$$R_{12} = \frac{M'_2}{L} \qquad R_{22} = \alpha R_{12}$$

$$R_{1n} = \frac{M'_n}{L} \qquad R_{2n} = \alpha R_{1n}$$

$$\begin{aligned}
M_{11} &= X'_{11} & M_{21} &= X''_{11} + X'_{21} \\
M_{12} &= X'_{12} - M_{11} & M_{22} &= X''_{12} + X'_{22} - M_{21} \\
M_{1n} &= X'_{1n} - M_{1,n-1} & M_{2n} &= X''_{1n} + X'_{2n} - M_{n,n-1}
\end{aligned} \qquad (3\text{-}19)$$

$$N_{11} = W_1 - H_{11} \qquad N_{21} = \frac{1}{2} W_1$$

$$N_{12} = W_2 - H_{12} \qquad N_{22} = \frac{1}{2} W_2$$

$$N_{1n} = W_n - H_{1n} \qquad N_{2n} = \frac{1}{2} W_n$$

$$H_{11} = \frac{M_{11}}{a_1} \qquad H_{21} = \frac{1}{2} W_1 - H_{11}$$

$$H_{12} = \frac{M_{12}}{a_1} \qquad H_{22} = \frac{1}{2} \sum_{1}^{2} W - H_{12}$$

$$H_{1n} = \frac{M_{1n}}{a_1} \qquad H_{2n} = \frac{1}{2} \sum_{1}^{n} W - H_{1n}$$

Multistory Frames

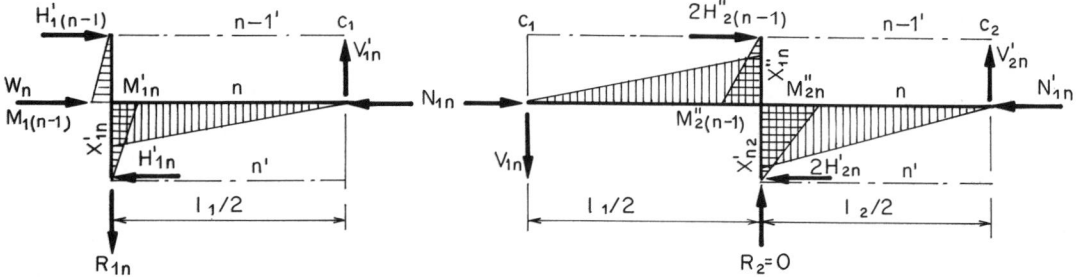

Figure 3-14(b)

Example

For illustration of the zero-point method, four floors from the top (1) downward of a three-bay frame—as illustrated in Fig. 3-15a—will be analyzed.

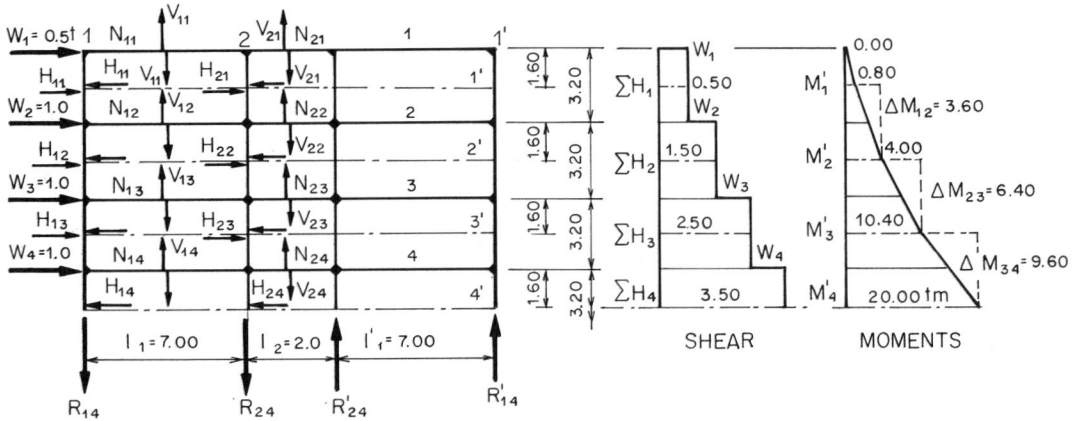

Figure 3-15(a)

$$W_1 = 0.5 \text{ t} \qquad W_2 \cdots W_4 = 1.0 \text{ t}$$

$$L_1 = L_1' = 7.00 \text{ m} \qquad L_2 = 2.00 \text{ m}$$

$$h_1 \cdots h_4 = 3.20 \text{ m} \qquad a_1 \cdots a_4 = \frac{1}{2}h = 1.60 \text{ m}$$

$$-R_{14} = R_{14}' \qquad -R_{24} = R_{24}'$$

$$R_1 = \alpha R_2$$

$$\alpha \cong \frac{L_1 I_2}{2 L_2 I_1} = \frac{7.00 \cdot 1.1}{2 \cdot 2.00 \cdot 1.0} = \sim 2$$

$$\overline{L} = 2L_1 + L_2(1 + \alpha) = 2 \cdot 7.00 + 2.00(1 + 2) = 20.00 \text{ m}$$

Shear, moments, and increase of moments are computed by Eqs. 3-16, 3-17, 3-18, and 3-19 and plotted in Fig. 3-15b.

$$R_{1n} = \frac{M'_n}{\overline{L}} \qquad R_{2n} = \alpha R_{1n}$$

$$R_{11} = \frac{M'_{11}}{\overline{L}} = \frac{0.80}{20.00} = 0.04 \text{ t} \qquad R_{21} = 2 \cdot R_{11} = 0.08 \text{ t}$$

$$R_{12} = \frac{M'_{12}}{\overline{L}} = \frac{4.00}{20.00} = 0.20 \text{ t} \qquad R_{22} = 2 \cdot R_{12} = 0.40 \text{ t}$$

$$R_{13} = \frac{M'_{13}}{\overline{L}} = \frac{10.40}{20.00} = 0.52 \text{ t} \qquad R_{23} = 2 \cdot R_{13} = 1.04 \text{ t}$$

$$R_{14} = \frac{M'_{14}}{\overline{L}} = \frac{20.00}{20.00} = 1.00 \text{ t} \qquad R_{24} = 2 \cdot R_{14} = 2.00 \text{ t}$$

$$V_{1n} = \frac{\Delta M_{n,n+1}}{\overline{L}} \qquad V_{2n} = (1 + \alpha)V_{1n}$$

$$V_{11} = \frac{M'_1}{\overline{L}} = \frac{0.80}{20.00} = 0.04 \text{ t} \qquad V_{21} = (1 + 2)\, 0.04 = 0.12 \text{ t}$$

$$V_{12} = \frac{\Delta M_{12}}{\overline{L}} = \frac{3.60}{20.00} = 0.18 \text{ t} \qquad V_{22} = 3V_{12} = 0.54 \text{ t}$$

$$V_{13} = \frac{\Delta M_{23}}{\overline{L}} = \frac{6.40}{20.00} = 0.32 \text{ t} \qquad V_{23} = 3V_{13} = 0.96 \text{ t}$$

$$V_{14} = \frac{\Delta M_{34}}{\overline{L}} = \frac{9.60}{20.00} = 0.48 \text{ t} \qquad V_{24} = 3V_{14} = 1.44 \text{ t}$$

$$X'_{1n} = \frac{L_1}{2} V_{1n} = -X''_{1n} \qquad X'_{2n} = \frac{L_2}{2} V_{2n} = -X''_{2n}$$

$$X'_{11} = \frac{L_1}{2} V_{11} = \frac{7.00}{2}\, 0.04 = 0.14 \text{ tm} \qquad X'_{21} = \frac{L_2}{2} V_{21} = \frac{2.00}{2} \cdot 0.12 = 0.12 \text{ tm}$$

$$X'_{12} = 3.50\, V_{12} = 0.63 \text{ tm} \qquad X'_{22} = 1.0\, V_{22} = 0.54 \text{ tm}$$

$$X'_{13} = 3.50\, V_{13} = 1.12 \text{ tm} \qquad X'_{23} = 1.0\, V_{23} = 0.96 \text{ tm}$$

$$X'_{14} = 3.50\, V_{14} = 1.68 \text{ tm} \qquad X_{24} = 1.0\, V_{24} = 1.44 \text{ tm}$$

$$M_{1n} = X'_{1n} - M_{1,n-1}$$

$$M_{11} = X'_{11} = 0.14 \text{ tm}$$

$$M_{12} = X'_{12} - M_{11} = 0.63 - 0.14 = 0.49 \text{ tm}$$

$$M_{13} = X'_{13} - M_{12} = 1.12 - 0.49 = 0.71 \text{ tm}$$

$$M_{14} = X_{14} - M_{13} = 1.68 - 0.71 = 0.97 \text{ tm}$$

Multistory Frames

$$M_{2n} = X''_{1n} + X'_{2n} - M_{21}$$

$M_{21} = X''_{11} + X'_{21} = 0.14 + 0.12 = 0.26$ tm

$M_{22} = X''_{12} + X'_{22} - M_{21} = 0.63 + 0.54 - 0.26 = 0.91$ tm

$M_{23} = X''_{13} + X'_{23} - M_{22} = 1.12 + 0.96 - 0.91 = 1.17$ tm

$M_{24} = X''_{14} + X'_{24} - M_{23} = 1.68 + 1.44 - 1.17 = 1.95$ tm

$$H_{1n} = \frac{M_{1n}}{a_1}$$

$$H_{11} = \frac{M_{11}}{a_1} = \frac{0.14}{1.60} = 0.088 \text{ t}$$

$$H_{12} = \frac{M_{12}}{a_1} = \frac{0.49}{1.60} = 0.306 \text{ t}$$

$$H_{13} = \frac{M_{13}}{a_1} = \frac{0.71}{1.60} = 0.444 \text{ t}$$

$$H_{14} = \frac{M_{14}}{a_1} = \frac{0.97}{1.60} = 0.606 \text{ t}$$

$$H_{2n} = \frac{1}{2}\sum_{1}^{n} W - H_{1n}$$

$H_{21} = \frac{1}{2} W - H_{11} = \frac{1}{2} 0.50 - 0.088 = 0.162$ t

$H_{22} = \frac{1}{2}\sum_{1}^{2} W - H_{12} = 0.75 - 0.306 = 0.444$ t

$H_{23} = \frac{1}{2}\sum_{1}^{3} W - H_{13} = 1.25 - 0.444 = 0.806$ t

$H_{24} = \frac{1}{2}\sum_{1}^{4} W - H_{14} = 1.75 - 0.606 = 1.144$ t

$$N_{1n} = W_n + H_{1(n-1)}$$

$N_{11} = W_1 - H_{11} = 0.50 - 0.088 = 0.412$ t

$N_{12} = W_2 + H_{11} - H_{12} = 1.00 + 0.088 - 0.306 = 0.782$ t

$N_{13} = W_3 + H_{12} - H_{13} = 1.00 + 0.306 - 0.444 = 0.862$ t

$N_{14} = W_4 + H_{13} - H_{14} = 1.00 + 0.444 - 0.606 = 0.838$ t

$$N_{2n} = \frac{1}{2} W_n$$

$$N_{21} = \frac{1}{2} W_1 = 0.25 \text{ t}$$

$$N_{22} = \frac{1}{2} W_2 = 0.50 \text{ t}$$

$$N_{23} = \frac{1}{2} W_3 = 0.50 \text{ t}$$

$$N_{24} = \frac{1}{2} W_4 = 0.50 \text{ t}$$

$$N'_{11} = N_{11} - N_{21} = 0.412 - 0.25 = 0.162 \text{ t}$$
$$N'_{12} = N_{12} - N_{22} = 0.782 - 0.50 = 0.282 \text{ t}$$
$$N'_{13} = N_{13} - N_{23} = 0.862 - 0.50 = 0.362 \text{ t}$$
$$N'_{14} = N_{14} - N_{24} = 0.838 - 0.50 = 0.338 \text{ t}$$

Checking:

$$2(H_{14} + H_{24}) - \Sigma W_4 = 2(0.606 + 1.144) - 3.5 = 0, (\Sigma X = 0)$$

$$V_{11} + V_{12} + V_{13} + V_{14} - R_{14}$$
$$= 0.04 + 0.18 + 0.32 + 0.48 - 1.00 = 0, (\Sigma Z = 0)$$

$$N_{11} + N_{12} + N_{13} + N_{14} + H_{14} - \sum_1^4 W$$
$$= 0.412 + 0.782 + 0.862 + 0.838 + 0.606 - 3.50 = 0, (\Sigma X = 0)$$

$$V_{21} + V_{22} + V_{23} + V_{24} - (R_{14} + R_{24})$$
$$= 0.12 + 0.54 + 0.96 + 1.44 - (1.00 + 2.00) = 0 \ (\Sigma Z = 0)$$

$$N_{21} + N_{22} + N_{23} + N_{24} + H_{14} + H_{24} - \sum_1^4 W$$
$$= 0.25 + 3 \cdot 0.50 + 0.606 + 1.144 - 3.50 = 0$$

The moments (M, X), shears (V, H), reactions (R), and normal forces (N) for span 4 are plotted in Fig. 3-15b.

Lateral Deflection

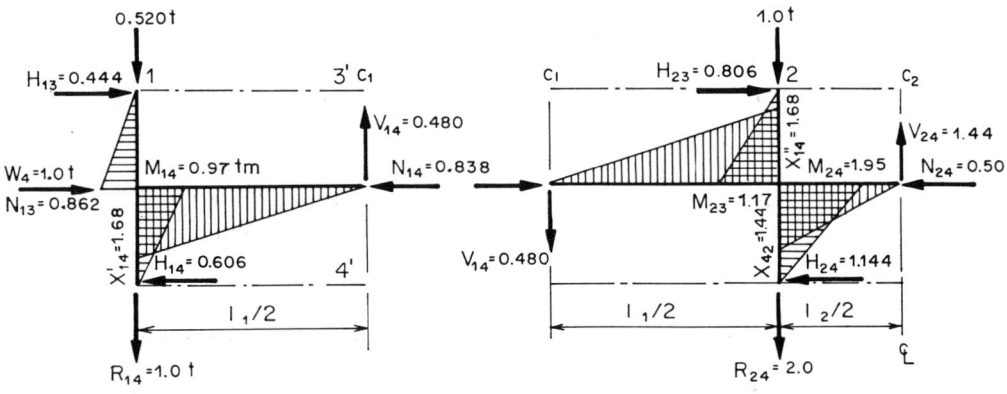

Figure 3-15(b)

LATERAL DEFLECTION

Horizontal deflection of a floor (u_n) due to loading (W_n) may be computed approximately by the assumption that all columns in a story have equal displacement and the zero points in columns are $h_n/2$. If this is not the case, average values have to be used.

If the floor beams are completely rigid, the lateral displacement (u'_n) of the story composed of two cantilevers would be

$$EI_{cn}u'_n = 2\frac{1}{3} H_n \left(\frac{h_n}{2}\right)^3 = \frac{1}{12} H_n h_n^3 \qquad (3\text{-}20)$$

However, the floor beams are not infinitely rigid, but deflect (Fig. 3-16).

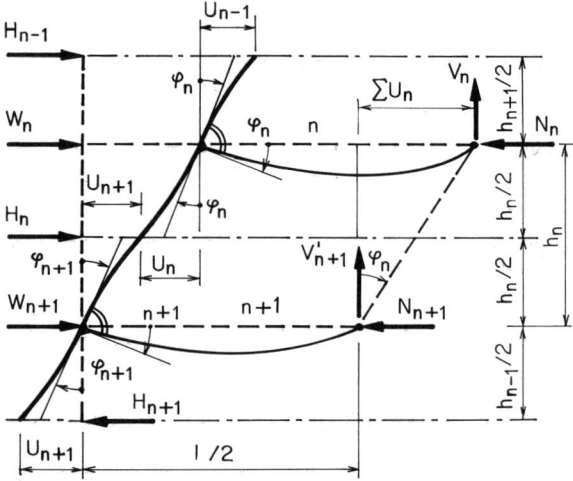

Figure 3-16

The moments of a beam composed also of two cantilevers subjected to the vertical loads V_n and V_{n+1} are:

$$M_n = \frac{L}{2} V_n, \qquad M_{n+1} = \frac{L}{2} V_{n+1} \qquad (3\text{-}21)$$

and the end rotations (φ_n and φ_{n+1}) for constant moment of inertia are

$$I_n = I_{n+1} = I \text{ (Eq. 1-17)}$$

$$EI\varphi_n = \frac{1}{3} V_n \left(\frac{L}{2}\right)^2 = \frac{1}{12} V_n L^2 \qquad (3\text{-}22)$$

$$EI\varphi_{n+1} = \frac{1}{3} V_{n+1} \left(\frac{L}{2}\right)^2 = \frac{1}{12} V_{n+1} L^2$$

The end rotation of the adjacent floor beams results in story displacement:

$$u_n'' = \frac{1}{E}\left(\frac{\varphi_n + \varphi_{n+1}}{I}\right)\frac{h_n}{2} = \frac{1}{12E}\left(\frac{V_n + V_{n+1}}{I} L^2 \frac{h}{2}\right) \qquad (3\text{-}23)$$

Thus, the total story displacement, illustrated in Fig. 3-16, is

$$u_n = u_n' + u_n'' = \frac{1}{12E}\left(\frac{H_n}{I_{cn}} h_n^3 + \frac{V_n + V_{n+1}}{I} L^2 \frac{h}{2}\right) \qquad (3\text{-}24)$$

The lateral deflection of the top floor (u_T) is obtained by adding the story deflections starting from the foundation (F). Thus,

$$u_T = \sum_F^1 u_n \qquad (3\text{-}25)$$

Example

Floor $n = 3$:

$$h_3 = 3.20 \text{ m}, L_3 = L_4 = L = 7.00 \text{ m}$$

$$V_{13} = 0.320 \text{ t}, V_{14} = 0.480 \text{ t}$$

$$H_{13} = 0.488 \text{ t}, H_{14} = 0.606 \text{ t}$$

$$E_c = 270000 \text{ kg/cm}^2 = 270 \text{ t/cm}^2$$

$$I_{3,4} = 320000 = 0.320 \cdot 10^6 \text{ cm}^4$$

$$I_{c3} = 160000 = 0.160 \cdot 10^6 \text{ cm}^4$$

In accordance with Eqs. 3-22 to 3-25,

$$EI_3 \varphi_3 = \frac{1}{12} V_{13} L_3^2 = \frac{1}{12} 0.320 \cdot 7.00^2 = 1.307 \text{ tm}^2$$

One-Story Multibay Frames

$$EI_4 \, \varphi_4 = \frac{1}{12} \cdot V_{14} \cdot L_4^2 = \frac{1}{12} 0.480 \cdot 7.00^2 = 1.960 \text{ tm}^2$$

$$EI_{c3} \, U_3' = \frac{1}{12} H_3 h_3^3 = \frac{1}{12} 0.488 \cdot 3.20^3 = 1.333 \text{ tm}^3 = 1.333 \cdot 10^6 \text{ tcm}^3$$

$$EI_{3,4} \, U_3'' = \frac{1}{12} (V_{13} + V_{14}) L^2 \cdot \frac{h_3}{2} = \frac{1}{12} (0.320 + 0.480) 7.0^2 \cdot 1.60 = 5.227 \text{ tm}^3$$

$$= (\varphi_3 + \varphi_4) \cdot \frac{h}{2} = (1.307 + 1.960) \frac{3.20}{2} = 5.227 \text{ tm}^3$$

$$u_3 = u_3' + u_3'' = \frac{1}{270} \left(\frac{1.333 \cdot 10^6}{0.160 \cdot 10^6} + \frac{5.227 \cdot 10^6}{0.320 \cdot 10^6} \right)$$

$$= \frac{1}{270} (8.331 + 16.334) = \sim 0.1 \text{ cm}.$$

The H- and V-values have been computed for $W = 1.0$ t. Thus, the true lateral deflection of floors must be multiplied by the actual W-values.

ONE-STORY MULTIBAY FRAMES

The determination of moments of one-story multibay frames is the same as for multistory, multibay frames, except that the s_u, d_u, a_u' and X_u', H_u–values in the Eqs. 3-1 to 3-9 are zero.

In construction of the multistory frames, the structural system changes from story to story (Fig. 3-17); therefore, only in its final state is the multistory frame theory valid.

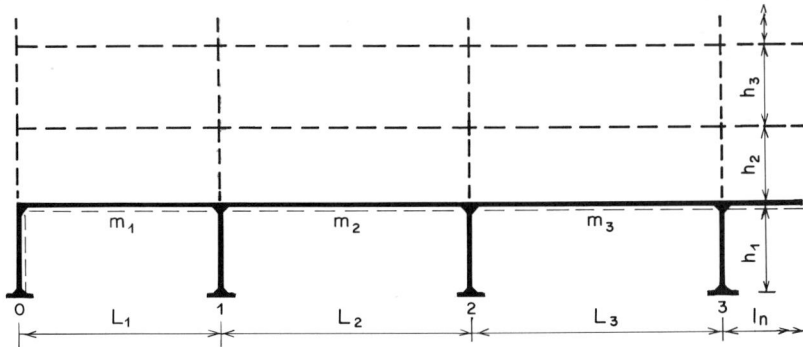

Figure 3-17

During construction, the structural system is acting as a one-story, multibay frame, which is less efficient than the system considered (Fig. 3-6). Also, the one-story frame is subjected to about twice as high loading and stresses as was used for the design. In addition, the temperature and shrinkage stresses mostly occur in the construction state, which may increase the flexural stresses considerably. For long spans and heavy, short columns, the temperature and shrinkage stresses can be high. Thus, critical stresses occur during construction and, if this fact is overlooked, may lead to considerable overstressing, cracking, and failures.

TEMPERATURE CHANGE AND SHRINKAGE

The coefficient of thermal expansion (αt) commonly is assumed to be 0.00001 for reinforced high-strength concrete. The change of temperature (ΔT) depends on climatic conditions—in the middle United States it is assumed to be $\pm 15°C$.

Thus, the change in length (ΔL) of unrestricted lateral displacement ($\Sigma \Delta L$) of a multistory parking structure (Fig. 3-17) with span

$$L_1 = L_2 = L_3 = 20 \text{ m}$$

$$\Delta L_2 = \alpha_t \, \Delta T L_2/2 = 1.0 \cdot 10^{-5} \cdot 15 \cdot 10 \cdot 100 = \pm 0.15 \text{ cm}$$

$$\Delta L_{1,2} = \alpha_t \, \Delta T (L_2/2 + L_1) = 1.0 \cdot 10^{-5} \cdot 15 \cdot 30 \cdot 100 = \pm 0.45 \text{ cm}$$

The shortening of length due to shrinkage commonly is considered equal to $-15°C$. Thus, the maximum possible shortening of the beam from the point of symmetry is

$$\Sigma \Delta L_{sr} = -\alpha_t \cdot \Delta T \cdot L = -1.0 \cdot 10^{-5} \cdot 15 \cdot 30 = -0.45 \text{ cm}$$

Under normal conditions, the ratio S_c/S_B is relatively small, resulting in rather high moments in beams and columns. The determination of these moments depends on the change of beam length (ΔL), rotation of the joints (φ), and lateral displacement (ψ_c), and is time-requiring and complex.

There is no simple method to determine these values. The most convenient method available is the deformation method.* It is based upon a geometrically determined principal system in which the redundants φ_J, φ_K, and ψ_c are zero. They are determined from the virtual work equation.

$$\delta_{AJ} = a_{JJ} \varphi_J + \Sigma a_{JK} \varphi_K + \Sigma a_{Jc} \psi_c + a_{J0} = 0$$

$$\delta_{Ac} = a_{cc} \psi_c + \Sigma a_{Jc} \varphi_J + \Sigma a_{Kc} \varphi_K + a_{c0} = 0 \qquad (3\text{-}26)$$

$$J = A, B \cdots K \qquad c = 1, 2, \cdots n$$

The a_{JJ}, a_{JK} and a_{Jc}-values in the first equation ($\delta A_J = 0$) are the virtual work of the joint moments in the principal system due to $\varphi_J = 1$, $\varphi_K = 1$ and $\psi_c = 1$. The a_{cc}- and a_{Jc}-values in the second equation ($\delta A_c = 0$) are the virtual work due to $\psi_c = 1$ and $\varphi_J = \varphi_K = 0$. The a_{J0}, a_{c0}-values are determined from the angular change of columns (ϑ) in the principal system.

*A. E. Komendant, D. E., *Contemporary Concrete Structures* (New York: McGraw-Hill Book Company, 1972).

Temperature Change and Shrinkage

To demonstrate this method, the moments (M_t) and thrust (H_t) of the first-floor frame (Fig. 3-17) of a parking structure for temperature rise $\Delta T = 15°$ and linear temperature expansion $\alpha_t = 1.0 \cdot 10^{-5}$ will be computed.

Since the lateral displacement at the symmetry line $\Delta L = 0$, the $\psi_c = 0$ and the two equations for joints 1 and 2 of the virtual work to determine the redundants φ_1 and φ_2 become

$$a_{22}\,\varphi_2 + a_{21}\,\varphi_1 = a_{20}$$

$$a_{11}\,\varphi_1 + a_{12}\,\varphi_2 = a_{10}$$

$$a_{22} = -\left(2\frac{4}{L_1'} + \frac{4}{h_1'}\right) = -\left(2\cdot\frac{4}{20.0} + \frac{4}{5.6}\right) = -1.114$$

$$h' = h\frac{I_B}{I_c} = 4.0\,\frac{0.021}{0.015} = 5.6 \text{ m}$$

$$a_{12} = -\frac{2}{L_1'} = -\frac{2}{20.0} = -0.10 \qquad L_1' = L_2' = L\frac{I_B}{I_B} = 20.0 \text{ m}$$

$$a_{11} = -\left(\frac{4}{L_1'} + \frac{4}{h'}\right) = -\left(\frac{4}{20.0} + \frac{4}{5.6}\right) = -0.914 \qquad a_{12} = a_{21}$$

$$n = 10 \qquad E_c = \frac{E_S}{n} = \frac{2100000}{10} = 21.0\cdot 10^5 \text{ t/m}^2$$

$$E_c I_B\,\alpha_t \Delta T = 21.0\cdot 10^5 \cdot 0.021 \cdot 1.0 \cdot 10^{-5} \cdot 15 = 6.615 \text{ tm}^2$$

$$\vartheta_2 = E_c I_B\,\alpha_t\,\Delta T\,\frac{L_2/2}{h} = 6.615\cdot\frac{10}{4.0} = 16.54$$

$$\vartheta_1 = E_c I_B\,\alpha_t \Delta T\,\frac{L_1 + L_2/2}{h} = 6.615\cdot\frac{30.0}{4.0} = 49.61$$

$$a_{20} = \frac{6\vartheta_2}{h'} = \frac{6\cdot 16.54}{5.6} = 17.72 \qquad a_{10} = \frac{6\cdot\vartheta_1}{h'} = \frac{6\cdot 49.61}{5.6} = 53.16$$

Thus

$$-1.114\,\varphi_2 - 0.100\,\varphi_1 = -17.72$$

$$-0.914\,\varphi_1 - 0.100\,\varphi_2 = -53.16$$

Solving these two linear equations, we obtain

$$\varphi_1 = -56.96 \qquad \varphi_2 = -10.08$$

Moments:

$$M_1' = -\frac{2}{L_1'}(2\cdot\varphi_1 + \varphi_2) = -\frac{2}{20.0}(2\cdot 56.96 + 10.08) = -12.40 \text{ tm}$$

$$M''_{c1} = \frac{2}{h'}(2 \cdot \varphi_1 - 3\vartheta_1) = \frac{2}{5.6}(2 \cdot 56.96 - 3 \cdot 49.61) = 12.47 \text{ tm}$$

$$M''_1 = -\frac{2}{L'_2}(2 \cdot \varphi_2 + \varphi_1) = \frac{2}{20.0}(2 \cdot 10.08 + 56.96) = 7.70 \text{ tm}$$

$$M''_{c2} = \frac{2}{h'}(2 \cdot \varphi_2 - 3\vartheta_2) = \frac{2}{5.6}(2 \cdot 10.08 - 3 \cdot 16.54) \cong 10.60 \text{ tm}$$

$$M'_2 = -\frac{2}{L'_2}(2 \cdot \varphi_2 + \varphi_2) = -\frac{2}{20.0}(2 \cdot 10.08 + 10.08) = -3.02 \text{ tm}$$

Checking:

$$\Sigma M_1 = M'_1 + M''_{c1} = -12.40 + 12.47 = +0.07 \text{ tm} \sim 0.0$$

$$\Sigma M_2 = M''_1 + M''_{c2} + M'_2 = -(7.70 + 3.02) + 10.60 = -0.12 \sim 0.0$$

Thrust: In accordance with practical experience, there are cold joints between the columns and foundations. Therefore, the columns are assumed to be only about 75 percent constrained.

Thus,

$$z''_c = 0.4 h = 0.4 \cdot 4.0 = 1.60 \text{ m}$$

$$z'_c = 0.6 h = 0.6 \cdot 4.0 = 2.40 \text{ m}$$

and the moments of columns at foundations are

$$H_1 = \frac{M''_{c1}}{z''_c} = \frac{12.47}{1.60} = 7.79 \text{ t}$$

$$H_2 = \frac{M''_{c2}}{z''_c} = \frac{10.60}{1.60} = 6.63 \text{ t}$$

$$M'_{c1} = H_1 z'_c = 7.79 \cdot 2.40 = 18.70 \text{ tm}$$

$$M'_{c2} = H_2 z'_c = 6.63 \cdot 2.40 = 15.90 \text{ tm}$$

Normal force:

$$N_1 = -H_1 = -7.79 \text{ t}$$

$$N_2 = -(H_1 + H_2) = -(7.79 + 6.63) = -14.42 \text{ t}$$

The moments are plotted in Fig. 3-18.

Due to low conductivity, concrete responds slowly to changes in temperature; therefore, the results obtained by Eq. 3-26 are conservative. However, in open parking structures in long-lasting low temperatures, the moments caused by shrinkage and a drop in temperature may occur at the same time and should not be overlooked.

Elastically Controlled Joints

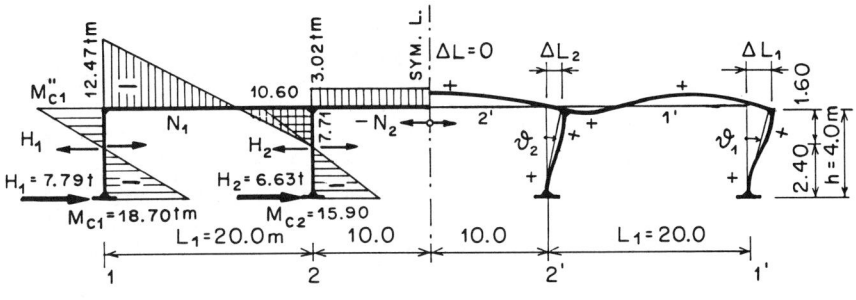

Figure 3-18

To reduce these moments considerably, elastically controlled joints between the columns and beams are recommended.

ELASTICALLY CONTROLLED JOINTS

The elastically controlled joints are obtained by introducing neoprene interfaces between beams and columns and providing sufficient elongation for reinforcing bars penetrating the neoprene. Such a joint is shown in Fig. 3-19.

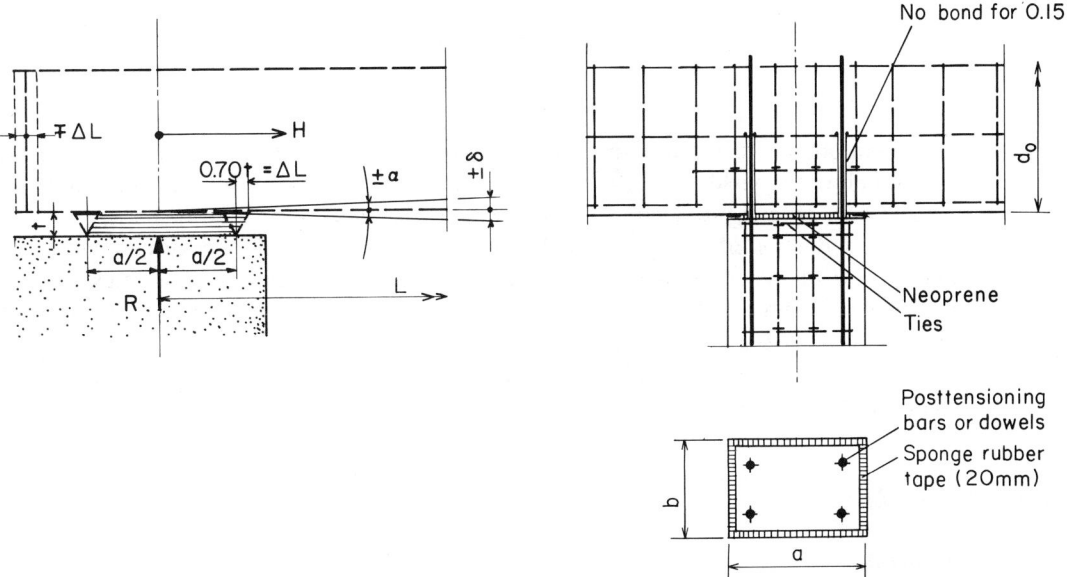

Figure 3-19

Under stress, the neoprene yields (Fig. 3-20*) and the stress in concrete is partially transferred into the reinforcing steel. The amount transferred depends on the area

*Tests carried out for Habitat construction, Montreal, Canada, 1967.

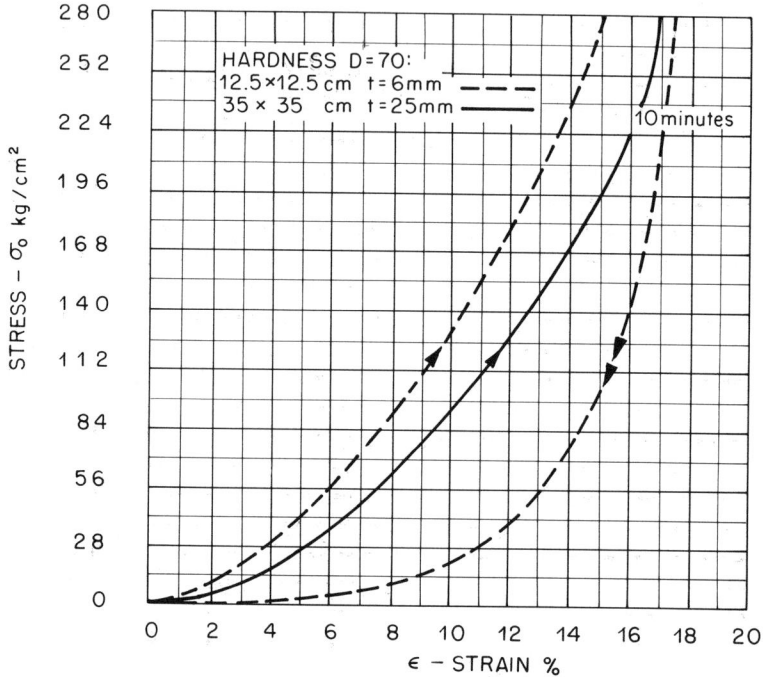

Figure 3-20

of the reinforcing steel. However, for most severe conditions the stress in the steel must remain below the yield point (0.75 σ_{UL}). Lateral displacement in this case depends on the thickness (t) and provided elongation length (Δh—nonbonded) of the steel.

The reduction of moments in columns is caused by reduced eccentricity and more efficient use of materials. Also, the stiffness of columns is reduced. One must realize that where eccentricity is zero, concrete strength is 100 percent used, but in cases of moments it is used about 15 percent.

Thus, the use of elastically controlled joints in frames results in slender columns, economy, and safety.

SIMPLE FRAMES

General

Structural analysis for simple frames also may be carried out by the loading factor method, as demonstrated in Fig. 3-21.

$$a_v = h/3 \qquad c_v = 0.5 \qquad d_v = \frac{S_c}{2 - c_v}$$

Simple Frames

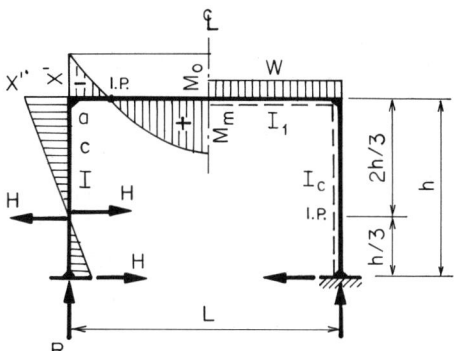

Figure 3-21

$$c' = c'' = \cfrac{1}{2 + \cfrac{S_1}{d_v}} = c$$

$$X'_1 = X''_1 = -c\,\frac{1-c}{1-c^2}\,\overline{L}$$

$$H = \frac{X'}{2h/3} \qquad M_a = H \cdot h/3 \qquad M_m = M_o - X'$$

For commonly used frames and loadings, the moments and reactions are given by the following closed formulas.*

Two-hinged frame

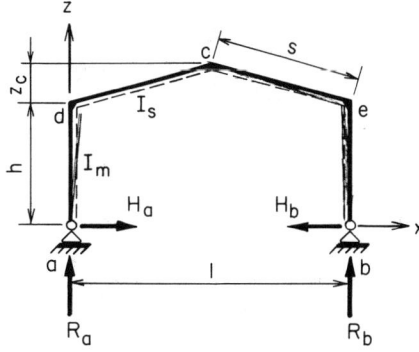

Figure 3-22(a)

---

*The formulas are taken from August Komendant, *Contemporary Concrete Structures* (McGraw-Hill Book Company, 1972).

$$\lambda = \frac{l}{h} \qquad \varphi = \frac{z_c}{h} \qquad \xi = \frac{x}{l} \qquad \zeta = \frac{z}{h}$$

$$\kappa = \frac{h}{s}\frac{I_s}{I_m}$$

$$\mu = 3 + \kappa + \varphi(3 + \varphi)$$

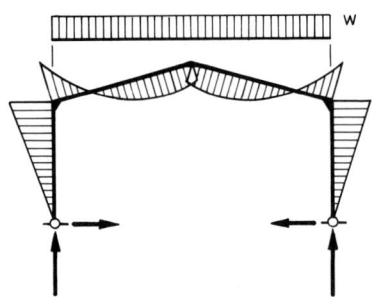

Figure 3-22(b)

$$R_a = R_b = \frac{wl}{2}$$

$$H_{a,b} = \frac{wl}{8}\lambda\frac{8 + 5\varphi}{4\mu} \tag{3-27}$$

$$M_{d,e} = -H_{a,b}h$$

$$M_c = \frac{wl^2}{8}\left[1 - (1 + \varphi)\frac{8 + 5\varphi}{4\mu}\right]$$

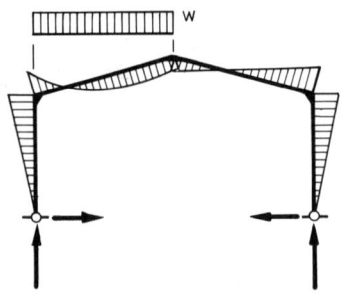

Figure 3-22(c)

$$R_a = \frac{3}{8}wl \qquad R_b = \frac{1}{8}wl \tag{3-28}$$

$$H_{a,b} = \frac{wl}{16}\lambda\frac{8 + 5\varphi}{4\mu}$$

Simple Frames

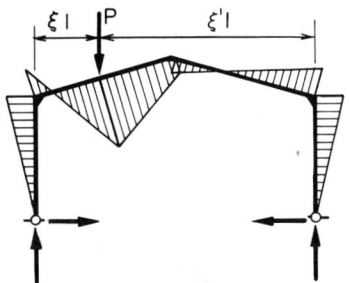

Figure 3-22(d)

$$R_a = \xi' P \qquad R_b = \xi P$$

$$H_{a,b} = \frac{P}{2} \lambda \frac{\xi}{\mu} [1.5(2 + \varphi) - \xi(3 + 2\varphi\xi)] \qquad (3\text{-}29)$$

$$M_{d,e} = -H_{a,b} h$$

$$M_c = \frac{Pl}{2} \left\{ \xi - (1 + \varphi) \frac{\xi}{\mu} [1.5(2 + \varphi) - \xi(3 + 2\varphi\xi)] \right\}$$

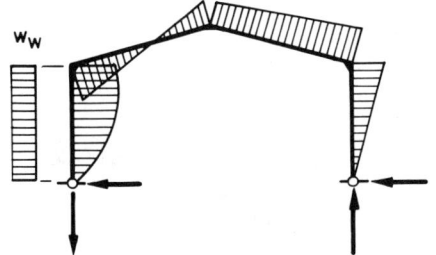

Figure 3-22(e)

$$-R_a = R_b = \frac{wh}{2} \qquad (3\text{-}30)$$

$$H_{a,b} = -\frac{w_w h^2}{2l} \left[1 \pm 1 - \frac{6(2 + \varphi) + 5\kappa}{8\mu} \right]$$

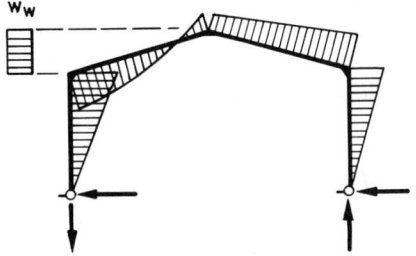

Figure 3-22(f)

$$-R_a = R_b = w_w z_c \frac{2h + z_c}{2l} \qquad (3\text{-}31)$$

$$H_{a,b} = -\frac{w_w z_c}{2}\left[\frac{\varphi}{8\mu}(4 + 3\varphi) \pm 1\right]$$

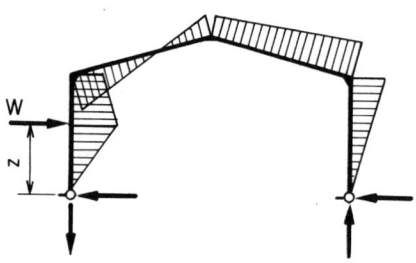

Figure 3-22(g)

$$-R_a = R_b = \frac{Wz}{l} \qquad (3\text{-}32)$$

$$H_{a,b} = -\frac{W}{2}\left[1 \pm 1 - \frac{\zeta}{2\mu}(6 + 3\varphi + 3\kappa - \kappa\zeta^2)\right]$$

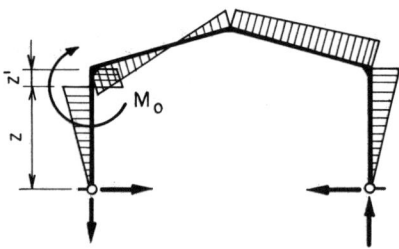

Figure 3-22(h)

$$-R_0 = R_b = \frac{M_0}{l}$$

$$H_{a,b} = \frac{3M_0}{4\mu h}[2 + \varphi + \kappa(1 - \zeta^2)] \qquad (3\text{-}33)$$

$$M_{d,e} = \frac{M_0}{2}\left[1 \pm 1 - \frac{3}{2\mu}(2 + \varphi + \kappa - \kappa\zeta^2)\right]$$

$$M_c = \frac{M_0}{2}\left[1 - \frac{3(1 + \varphi)}{2\mu}(2 + \varphi + \kappa - \kappa\zeta^2)\right]$$

Simple Frames

Temperature change:

$$R_a = R_b = 0 \tag{3-34}$$

$$H_{a,b} = \frac{3l}{2\mu s}\frac{EI_s}{h^2}\alpha_t \Delta T$$

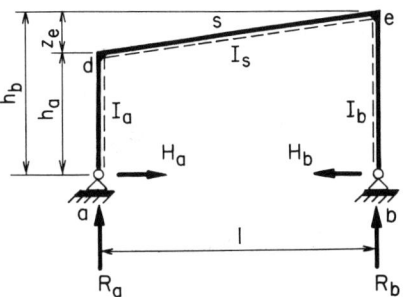

Figure 3-23(a)

$$\lambda_a = \frac{h_a}{h_b} \qquad \lambda_b = \frac{h_b}{h_a} \qquad \varphi_a = \frac{z_e}{h_a} \qquad \varphi_b = \frac{z_e}{h_b}$$

$$\nu_a = \frac{l}{h_a} \qquad \nu_b = \frac{l}{h_b} \qquad \kappa_a = \frac{h_a}{s}\frac{I_s}{I_a} \qquad \kappa_b = \frac{h_b}{s}\frac{I_s}{I_b}$$

$$\mu = \lambda_a(1 + \kappa_a) + 1 + \lambda_b(1 + \kappa_b)$$

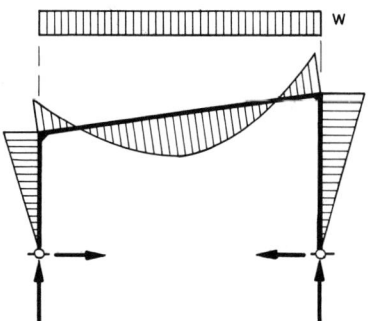

Figure 3-23(b)

$$R_a = R_b = \frac{wl}{2} \tag{3-35}$$

$$H_{a,b} = \frac{wl}{8\mu}(\nu_a + \nu_b)$$

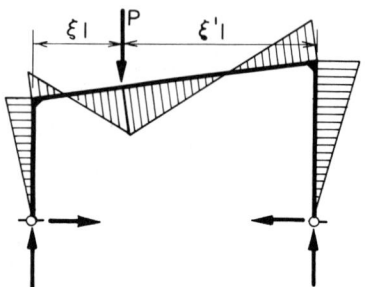

Figure 3-23(c)

$$R_a = \xi' P, \quad R_b = \xi P \tag{3-36}$$

$$H_{a,b} = \frac{P}{2\mu} [\nu_a(\xi - \xi^3) + \nu_b (\xi' - \xi'^3)]$$

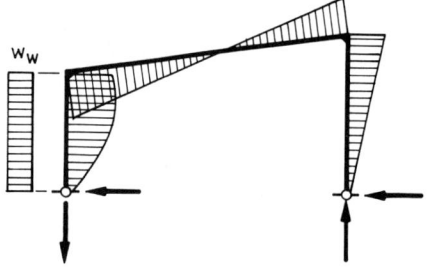

Figure 3-23(d)

$$-R_a = R_b = \frac{w_w h_a^2}{2\,l} \tag{3-37}$$

$$H_{a,b} = -\frac{w_w h_a}{2} \left\{ 1 \pm 1 - \frac{1}{4\mu} [2 + \lambda_a(4 + 5\kappa_a)] \right\}$$

Temperature change:

$$H_{a,b} = \frac{3}{\mu} \frac{l}{s} \frac{EI_s}{h_a^2} \alpha_t \Delta T \tag{3-38}$$

Fixed frames

$$\kappa = \frac{h}{l} \frac{I_c}{I_v} \qquad \xi = \frac{x}{l} \qquad \xi' = \frac{x'}{l}$$

$$\mu = 2 + \kappa \qquad \zeta = \frac{z}{h} \qquad \zeta' = \frac{z'}{h} \tag{3-39}$$

$$\nu = 1 + 6\kappa$$

Simple Frames

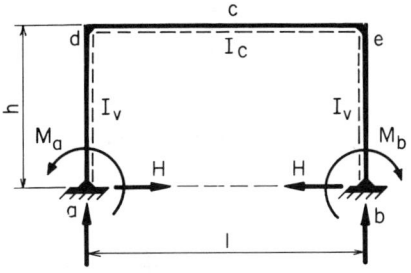

Figure 3-24(a)

$$R_a = R_b = \frac{wl}{2}$$

$$H_{a,b} = \frac{1}{4\mu} \frac{wl^2}{h}$$

$$M_{a,b} = \frac{wl^2}{12\mu}$$

$$M_{d,e} = -\frac{wl^2}{6\mu}$$

(3-40)

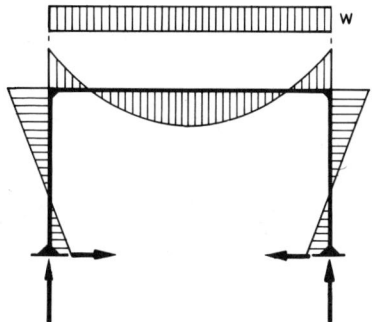

Figure 3-24(b)

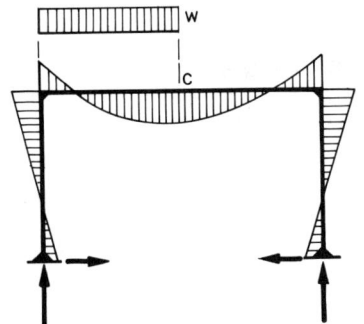

$$R_a = \frac{wl}{8}\left(3 + \frac{1}{4\nu}\right) \qquad R_b = \frac{wl}{8}\left(1 - \frac{1}{4\nu}\right)$$

$$H_{a,b} = \frac{1}{8\mu} \frac{wl^2}{h}$$

(3-41)

Figure 3-24(c)

$$M_{a,b} = \frac{wl^2}{24}\left(\frac{1}{\mu} \mp \frac{3}{8\nu}\right)$$

$$M_{d,e} = -\frac{wl^2}{24}\left(\frac{2}{\mu} \pm \frac{3}{8\nu}\right)$$

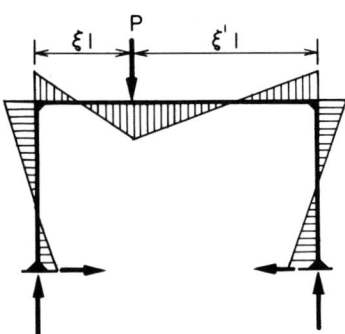

Figure 3-24(d)

$$R_a = P\xi'\left(1 + \frac{\xi(\xi' - \xi)}{\nu}\right) \qquad R_b = P - R_a$$

$$H_{a,b} = \frac{3}{2}\frac{Pl}{h}\frac{\xi\xi'}{\mu} \qquad\qquad\qquad\qquad\qquad (3\text{-}42)$$

$$M_{a,b} = \frac{Pl}{2}\xi\xi'\left[\frac{1}{\mu} \mp \frac{1}{\nu}(1 - 2\xi)\right]$$

$$M_{d,e} = -\frac{Pl}{2}\xi\xi'\left[\frac{2}{\mu} \pm \frac{1}{\nu}(1 - 2\xi)\right]$$

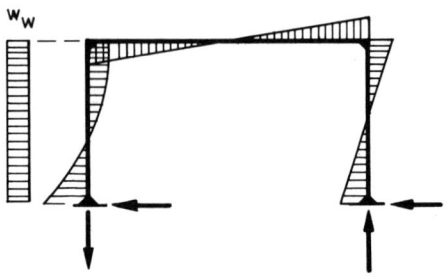

Figure 3-24(e)

Simple Frames

$$-R_a = R_b = \frac{w_w h^2}{l} \frac{\kappa}{\nu}$$

$$H_{a,b} = -\frac{w_w h}{4}\left(1 \pm 2 + \frac{1}{2\mu}\right)$$

$$M_{a,b} = -\frac{w_w h^2}{4}\left[\frac{3+\kappa}{6\mu} \pm \left(1 - \frac{2\kappa}{\nu}\right)\right] \quad (3\text{-}43)$$

$$M_{d,e} = -\frac{w_w h^2}{4}\kappa\left(\frac{1}{6\mu} \mp \frac{2}{\nu}\right)$$

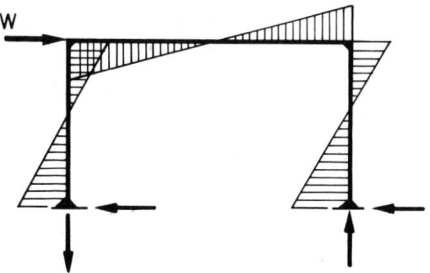

Figure 3-24(f)

$$-R_a = R_b = \frac{Wh}{l}$$

$$H_{a,b} = \mp\frac{W}{2}$$

$$M_{a,b} = \mp\frac{3}{2}Wh\left(\frac{1}{3} - \frac{\kappa}{\nu}\right) \quad (3\text{-}44)$$

$$M_{d,e} = \pm\frac{3}{2}Wh\frac{\kappa}{\nu}$$

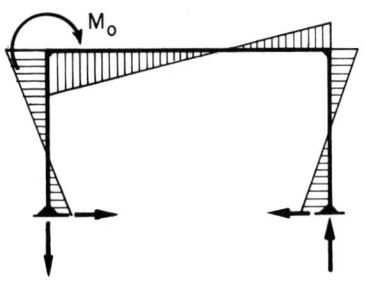

Figure 3-24(g)

$$-R_a = R_b = \frac{M_0}{l}\left(1 - \frac{1}{\nu}\right)$$

$$H_{a,b} = \frac{3}{2\mu}\frac{M_0}{h}$$

$$M_{a,b} = \frac{M_0}{2}\left(\frac{1}{\mu} \mp \frac{1}{\nu}\right)$$

$$M_{d,e} = \frac{M_0}{2}\kappa\left(\frac{1}{\mu} \pm \frac{6}{\nu}\right)$$

(3-45)

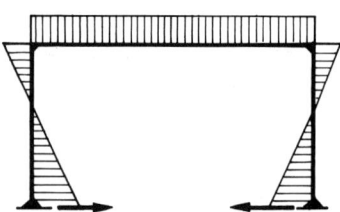

Figure 3-24(h)

$$H_{a,b} = \frac{2\kappa + 1}{\kappa}\frac{3}{\mu h^2} EI_c \alpha_t \Delta T$$

$$z_0 = h\left(1 - \frac{\kappa}{2\kappa + 1}\right) \qquad z_d = h - z_0$$

$$M_{a,b} = \frac{\kappa + 1}{\kappa}\frac{3}{\mu h} EI_c \alpha_t \Delta T$$

$$M_{d,e} = -\frac{3}{\mu h} EI_c \alpha_t \Delta T$$

(3-46)

4

Trusses

GENERAL

The basic geometry of trusses is triangular. The members of a truss—top (T) and bottom chords (B) and web members (verticals and diagonals)—are connected together into triangles at the joints. When triangular, no member of a truss can change its length unless the length of the other members is changed. Thus, the shape of the triangle is rigid and only two equations of equilibrium $\Sigma X = 0$ and $\Sigma Z = 0$ at each joint are required to determine the force in members. For statical analysis, the joints are assumed to be frictionless pins. However, this is not the case in the actual behavior of the truss.

This assumption is more or less justified for steel but not for concrete trusses. Therefore, most trusses are designed in steel where the bending stresses in members are of secondary order. In reinforced concrete trusses, the bending stresses may be higher than the normal stresses and, therefore, should not be overlooked.

SIMPLE TRUSSES

The trusses illustrated in Fig. 4-1a are suitable for reinforced or prestressed concrete.
The forces in members are determined by the method of sections. The members of a section are cut so that either side of the section is in equilibrium.

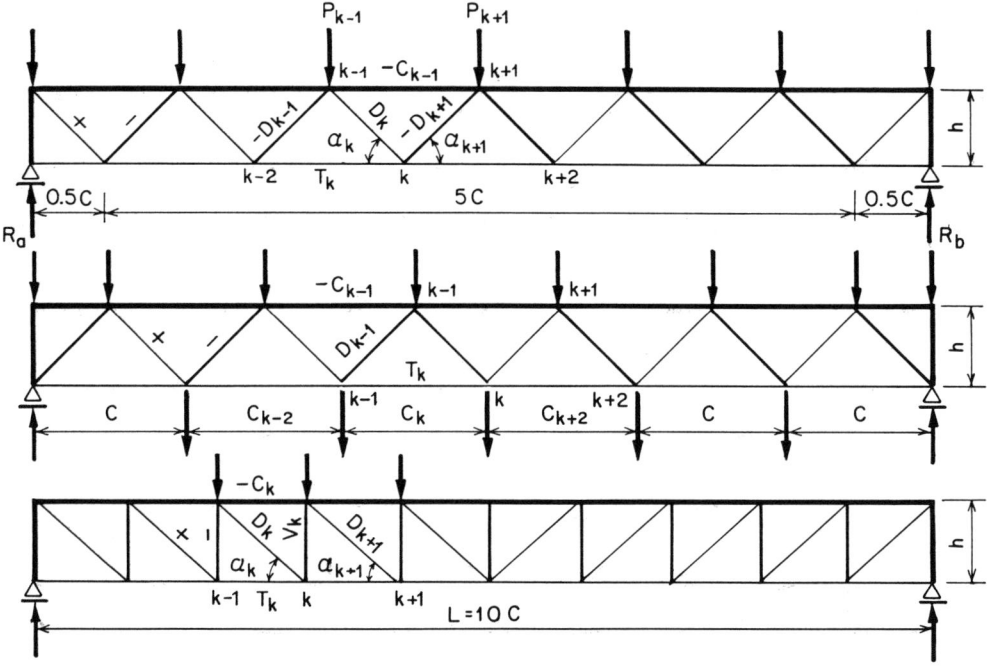

Figure 4-1(a)

First, the external moments M_{k-1}, M_k, M_{k+1}, due to concentrated loads at joints, are computed as for simply supported beams (Chapter 1) (Fig. 4-1b). Then the member forces and shear are determined by Eqs. 4-1, 4-2, and 4-3.

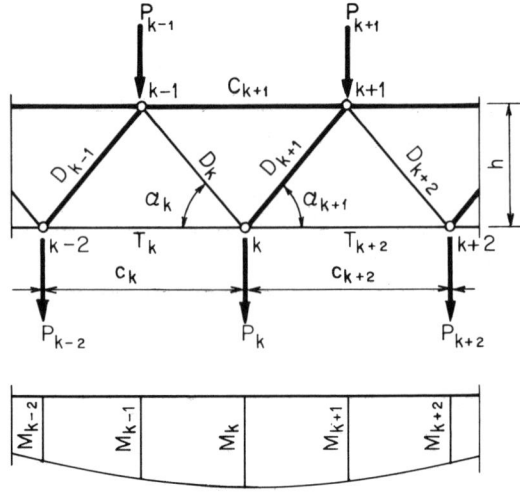

Figure 4-1(b) and (c)

Simple Trusses

$$T_k = \frac{M_{k-1}}{h}$$

$$C_{k+1} = -\frac{M_k}{h}$$

$$D_k = \frac{V_k}{\sin \alpha_k}$$

$$D_{k+1} = -\frac{V_{k+1}}{\sin \alpha_{k+1}}$$

(4-1)

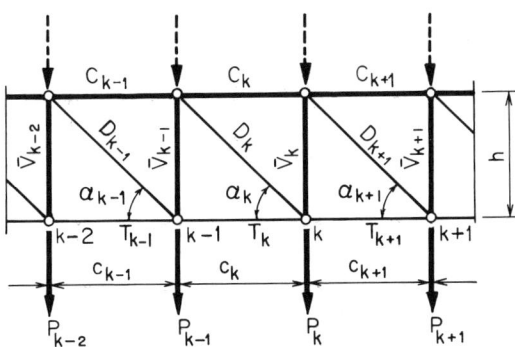

Figure 4-1(d)

$$T_k = \frac{M_k}{h} = -C_k$$

$$D_k = \frac{V_k}{\sin \alpha_k}$$

$$V_k = R - \sum_0^{k-1} P$$

$$\overline{V}_k = -V_k \pm P_k$$

(4-2)

The type of truss shown in Fig. 4-2 is obtained by overlapping the truss types in Figs. 4-1c,d and 4-1e. The member forces are obtained by superposition.

The bending stresses in members can be determined by the deformation method. However, the use of prestressing reduces the bending stresses to secondary stresses.

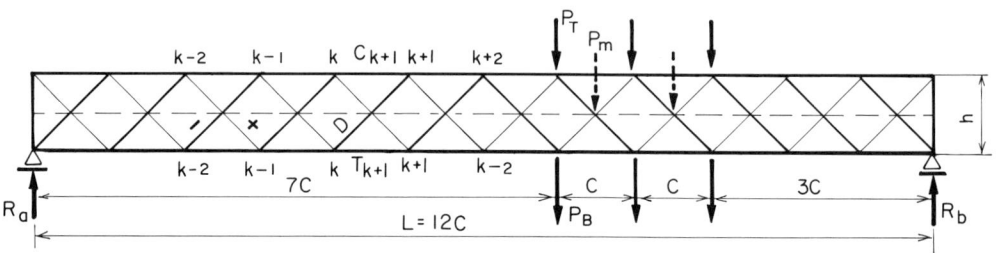

Figure 4-2

VIERENDEEL GIRDERS

Simplified Analysis

The Vierendeel girder is not a truss in the true sense because the characteristic triangular geometry of a true truss is missing. The carrying capacity of a Vierendeel is accomplished by chords and verticals only. Thus, it is more of a frame than a truss. A typical Vierendeel girder is shown in Fig. 4-3.

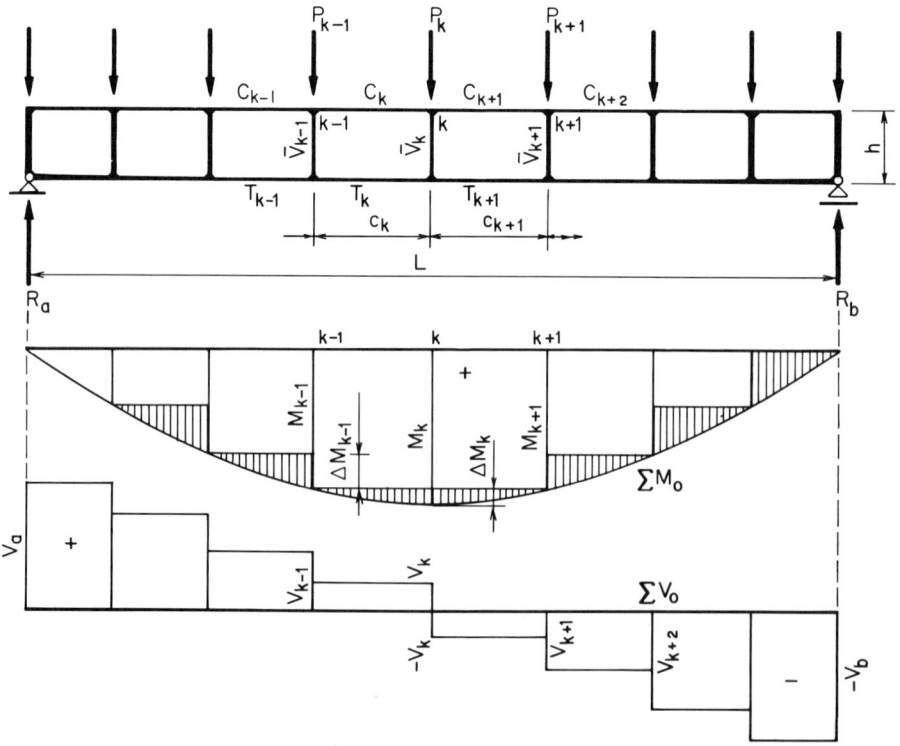

Figure 4-3

Vierendeel Girders

As the normal compressive force C_T in the top chord and normal tensil force T_B in bottom chord between the joints $k-1, k$ or $k, k+1$ and so on, are constant and equal ($\Sigma X = 0$), so also are the internal moments due to normal forces $-C_T z_0 = T_B z_0$ = constant. But the external moment M_k, due to loading, changes from joint to joint by a moment of ΔM_k. Thus, to obtain equilibrium, ΔM_k must be balanced by beam action in the chords.

Chords: $I_T = I_B$ If the moments of inertia of the top and bottom chords are equal ($I_T = I_B$), the zero points in the chords and verticals are $L/2, h/2$. The ΔM_k must be carried equally by the chord end moments

$$\sum_{B}^{T} M'_k = \sum_{B}^{T} M''_k$$

Thus,

$$M'_{kT} = M''_{kT} = M'_{kB} = M''_{kB} = \frac{\Delta M_k}{4} \qquad (4\text{-}3)$$

and the end moments of the verticals:

$$M''_{k,v} = M''_k + M'_{k+1} \qquad (4\text{-}4)$$

$$V_k = \frac{M''_{kv}}{c_k/2} \qquad V_{k,v} = \frac{M''_{kv}}{h/2}$$

The end moments of chords M_k, verticals $M_{k,v}$, and shears $V_k, V_{k,v}$ are illustrated in Fig. 4-4.

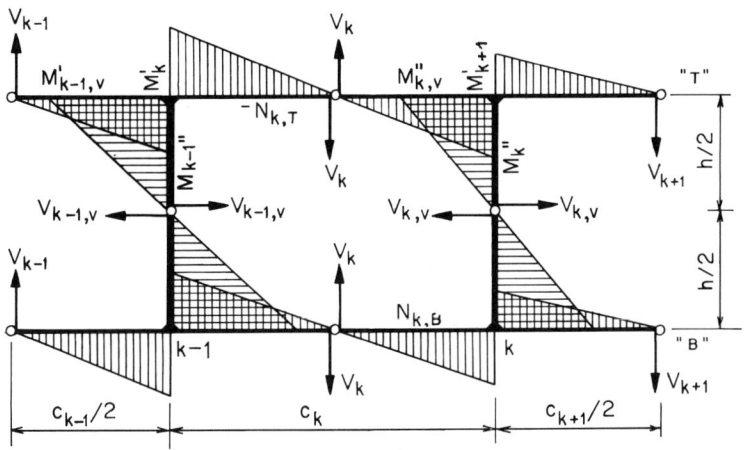

Figure 4-4

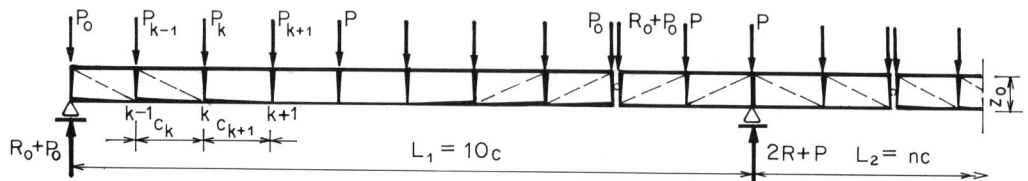
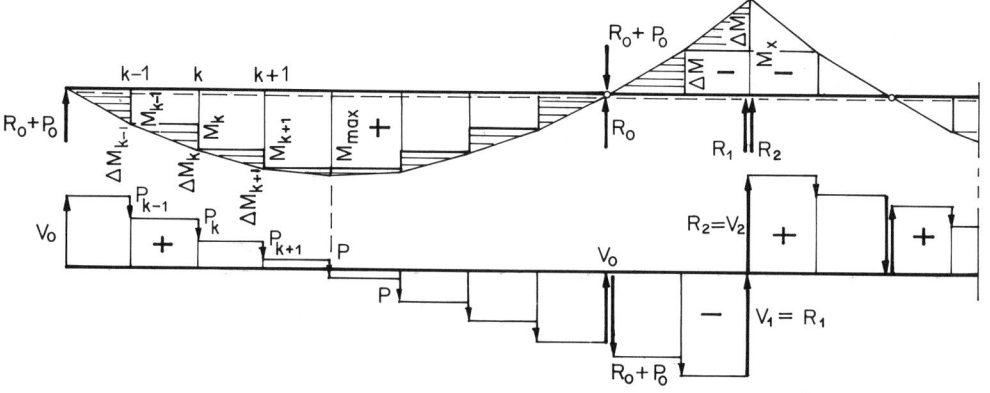

Figure 4-5

Continuous Gerber-type Vierendeels can be computed by the zero-point method. The distance (a''_1) of zero point from support 1 may be chosen for span 1 from condition $M_{m1} \sim M''_{1x}$. A continuous Vierendeel girder is shown in Fig. 4-5.

Chords: $I_T \neq I_B$ The case in which the moments of inertia of chords are not equal ($I_T \neq I_B$) will now be discussed.

The distribution of the $X'_k X''_k$ moments in chords occurs according to their stiffness. As the bottom chord is subject to tensile force (T), which has to be carried by steel, its cross section should be designed relatively small.

The total moment of inertia of the chords is:

$$I'_k = \sum_B^T I_k \qquad I''_k = \sum_B^T I_k$$

$$\sum_B^T X'_k = \frac{I'_k}{I'_k + I''_k} \Delta M_k \tag{4-5}$$

$$\sum_B^T X''_k = \frac{I''_k}{I'_k + I''_k} \Delta M_k$$

Zero-point distances:

$$a'_k = \frac{\sum_B^T X'_k + \sum_B^T X''_k}{c_k \sum_B^T X'_k} \qquad a''_k = \frac{\sum_B^T X'_k + \sum_B^T X''_k}{c_k \sum_B^T X''_k} \tag{4-6}$$

Vierendeel Girders

Normal forces:

$$-C_k = T_k = \frac{M_{k-1} + \sum_B^T X'_k}{z_0}$$

$$= \frac{M_k - \sum_B^T X''_k}{z_0}$$

(4-7)

Chord moments:

$$-X'_{kT} = \frac{I'_{kT}}{I'_{kT} + I'_{kB}} \sum_B^T X'_k \qquad -X'_{kB} = \frac{I'_{kB}}{I'_{kT} + I'_{kB}} \sum_B^T X'_k$$

$$X''_{kT} = \frac{I''_{kT}}{I''_{kT} + I''_{kB}} \sum_B^T X''_k \qquad X''_{kB} = \frac{I''_{kB}}{I''_{kT} + I''_{kB}} \sum_B^T X''_k$$

(4-8)

Moments in verticals ($\Sigma M = 0$):

$$M_{k-1,T} = X''_{k-1,T} + X'_{kT} \qquad M_{k-1,B} = X'_{k-1,B} + X'_{kB}$$

$$M_{kT} = X''_{kT} + X'_{k+1,T} \qquad M_{kB} = X''_{kB} + X'_{k+1,B}$$

(4-9)

Shears:

$$V_{kT} = \frac{X'_{kT}}{a'_k} \qquad V_{kB} = \frac{X'_{kB}}{a'_k} = \frac{X''_{kB}}{a''_k}$$

$$V_{k+1} = \frac{M_{k-1,T}}{a_{k-1}} \qquad V_k = \frac{M_{kT}}{a_k}$$

(4-10)

The moments and shear forces in chords and verticals are plotted in Fig. 4-6.

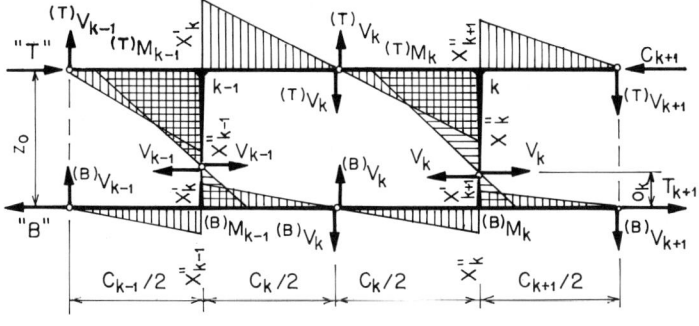

Figure 4-6

The economy of Vierendeels can be considerably improved by the use of diagonals in end bays and by prestressing. This is especially true for longer spans.

The deflection of Vierendeels can be roughly estimated in the same way as for beams by adding the deflections $u = \delta$, (Eq. 3-7) due to horizontal load, as computed for a vertical one-bay frame.

Example

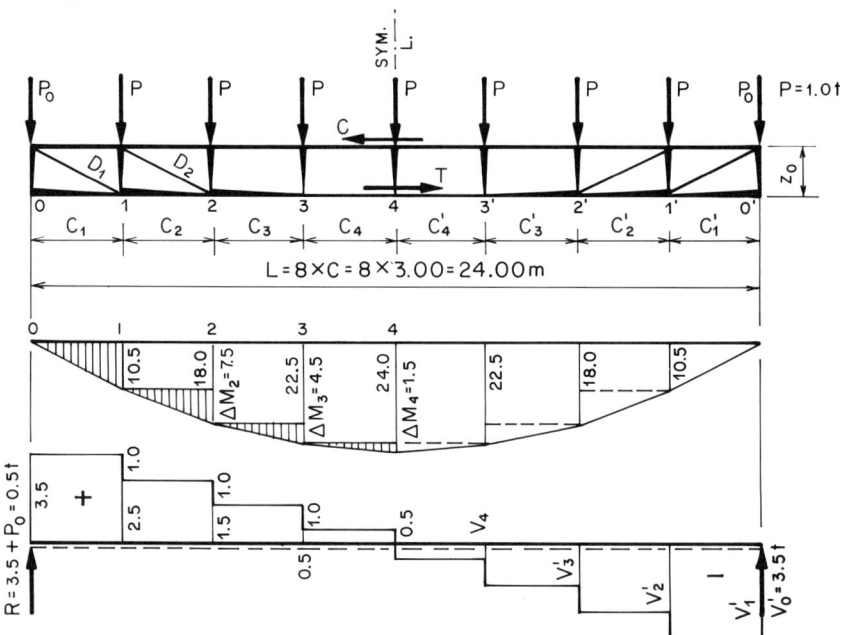

Figure 4-7

$P_0 = P_0' = P_1/2 = 1.0/2 = 0.5$ t

$P_1 = P_1 \ldots P_4 = 1.0$ t $\quad c = 3.0$ m

$R = 4.0$ t, $R' = 4.0 - 0.50 = 3.5$ t

$z_0 = 2.4$ m

$M_1 = 3.5 \cdot 3.0 = 10.5$ tm

$M_2 = 3.5 \cdot 6.0 - 1.0 \cdot 3.0 = 18.0$ tm

$M_3 = 3.5 \cdot 9.0 - 1.0(6.0 + 3.0) = 22.5$ tm

$M_4 = 3.5 \cdot 12.0 - 1.0(9.0 + 6.0 + 3.0) = 24.0$ tm

$V_1 = R - P_0 = 4.0 - 0.5 = 3.5$ t $= V_0$

$V_2 = V_1 - P_1 = 3.5 - 1.0 = 2.5$ t

Vierendeel Girders

$$V_3 = V_2 - P_2 = 2.5 - 1.0 = 1.5 \text{ t}$$
$$V_4 = V_3 - P_3 = 1.5 - 1.0 = 0.5 \text{ t}$$

Spacing of Vierendeels:

$= 10.00 \, m$

$\Sigma w = w_D + w_L \cong 5.0 \text{ t/m}$

$P = c\Sigma w = 3.00 \cdot 5.0 = 15 \text{ t}$

$ + D.L.\text{girder} \sim \underline{3 \text{ t}}$
$ P = 18 \text{ t}$

Actual loads:

$\Delta M_1 = M_1 - 0 = 10.5 \text{ tm}$

$\Delta M_2 = M_2 - M_1 = 18.0 - 10.5 = 7.5 \text{ tm}$

$\Delta M_3 = M_3 - M_2 = 22.5 - 18.0 = 4.5 \text{ tm}$

$\Delta M_4 = M_4 - M_3 = 24.0 - 22.5 = 1.5 \text{ tm}$

$M_{4,\max} = 18 \cdot 24.0 = 180 \text{ tm}$

Sectional coefficients:

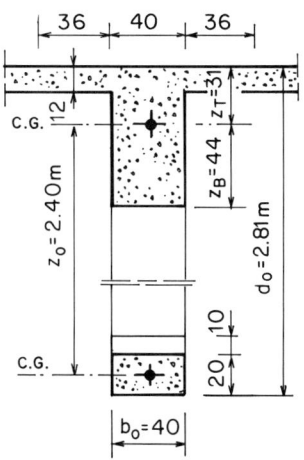

Figure 4-8

Top chord: $(b' = 2 \cdot 3.6 = 7.2 \text{ dm})$ (dm = 10 cm)

$$\begin{aligned} A_T &= 4.0 \cdot 7.5 = 30.00 \text{ dm}^2 \\ &+ 7.2 \cdot 1.2 = 8.65 \text{ dm}^2 \end{aligned} \bigg| = 38.65 \text{ dm}^2$$

$$\begin{aligned} Q &= 30.00 \cdot 3.75 = 112.50 \text{ dm}^3 \\ &+ 8.65 \cdot 0.60 = 5.19 \text{ dm}^3 \end{aligned} \bigg| = 117.69 \text{ dm}^3$$

$$z_T = \frac{Q}{A_T} = \frac{117.69}{38.65} = 3.05 \text{ dm} \sim 3.1 \text{ dm}, \; z_B = 4.4 \text{ dm}$$

$$I_{TT} = \frac{30.0 \cdot 7.5^2}{3} + \frac{8.65 \cdot 0.12^2}{3} = 567 \text{ dm}^4$$

$$I_{T0} = I_{TT} - z_T^2 A_T = 567 - 3.1^2 \cdot 38.65 = 195 \text{ dm}^4$$

Bottom chord:

$$d_0 = 2.0 \text{ dm } b'_0 = 4.0 - A_S \sim 3.0 \text{ dm}$$

$$A''_B = 2.0 \cdot 3.0 = 6.0 \text{ dm}^2 \quad A'_B = 3.0 \cdot 3.0 = 9.0 \text{ dm}^2$$

$$I''_{B0} = \frac{6.0 \cdot 2.0^2}{12} = 2.0 \text{ dm}^4 \quad I'_{B0} = \frac{9.0 \cdot 3.0^2}{12} \cong 6.8 \text{ dm}^4$$

$$I'_{B4} = I''_{B4}$$

Vierendeel—no diagonals. $-C = T$—constant along each bay; therefore ΔM must be balanced by the moments of four ends of the chords in accordance with the rigidity.

Thus,

$$\Delta M_k = M'_{Tk} + M'_{Bk} + M''_{Tk} + M''_{Bk} \quad P = 18 \text{ t}$$

$$\Sigma I_k = 2 \cdot I_{T0} + I'_{B0} + I''_{B0} = 2 \cdot 195 + 6.8 + 2.0 = \sim 400 \text{ dm}^4$$

$$k_T = \frac{I_{T0}}{\Sigma I_k} = \frac{195}{400} = 0.487$$

$$k'_B = \frac{I'_{B0}}{\Sigma I_k} = \frac{6.8}{400} = 0.017 \qquad \Sigma k = 2 \cdot 0.487 + 0.017 + 0.005 = 0.996 \sim 1.0$$

$$k''_B = \frac{I''_{B0}}{\Sigma I_k} = \frac{2.0}{400} = 0.005$$

$$M'_{T1} = k_T \cdot \Delta M_{T1} = 0.487 \cdot 10.5 = 5.11 \text{ tm} = M''_{T1}$$

$$M'_{B1} = k'_B \cdot \Delta M_{T1} = 0.017 \cdot 10.5 = 0.18 \text{ tm} \qquad \Sigma M_1 = 2 \cdot 5.11 + 0.18 + 0.05 = 10.45 \text{ tm} \sim \Delta M_1$$

$$M''_{B1} = k''_B \cdot \Delta M_{T1} = 0.005 \cdot 10.5 = 0.05 \text{ tm}$$

$$M_{01} = 10.5 - (5.11 + 0.05) = 5.34 \text{ tm}, (M_0 = 0)$$

$$-C_1 = T_1 = \frac{PM_{01}}{z_0} = \frac{18 \cdot 5.34}{2.40} = 40 \text{ t}$$

$$M'_{T2} = 0.487 \cdot 7.5 = 3.65 \text{ tm} = M''_{T2}$$

$$M'_{B2} = 0.017 \cdot 7.5 = 0.13 \text{ tm} \qquad \sum_B^T M_2 = 2 \cdot 3.65 + 0.13 + 0.04$$

$$M''_{B2} = 0.005 \cdot 7.5 = 0.04 \text{ tm} \qquad \sim = 7.5 \text{ tm} = \Delta M_2$$

Vierendeel Girders

$$M_{02} = M_1 + M'_{T2} + M'_{B2} = 10.5 + 3.65 + 0.13 = 14.28 \text{ tm}$$

$$-C_2 = T_2 = \frac{18 \cdot 14.28}{2.40} = 107 \text{ t}$$

$$M'_{T3} = 0.487 \cdot 4.5 = 2.20 \text{ tm} = M''_{T3}$$

$$M'_{B3} = 0.017 \cdot 4.5 = 0.08 \text{ tm}$$

$$M''_{B3} = 0.005 \cdot 4.5 = 0.02 \text{ tm}$$

$$\sum_{B}^{T} M_3 = \Delta M_3$$

$$M_{03} = 18 + 2.20 + 0.08 = 20.28 \text{ tm}$$

$$-C_3 = T_3 = \frac{18 \cdot 20.28}{2.40} = 152 \text{ t}$$

$$-C_4 = T_4 = \frac{18 \cdot 23.26}{2.40} = 175 \text{ t}$$

The moments are illustrated in Fig. 4-9a.

Reinforcing (see Chapter 6)

Top chord (critical only):

$$k_x = \frac{15\sigma_c}{15\sigma_c + \sigma_s} = \frac{15 \cdot 0.140}{15 \cdot 0.140 + 1.400} = 0.60$$

$$x = k_x d = 0.60 \cdot 70 = 42 \text{ cm}, \quad z_0 = \left(1 - \frac{k_x}{3}\right)d = \left(1 - \frac{0.60}{3}\right)$$

$$= 70 = 56 \text{ cm}$$

$$m = \frac{\sigma_s}{\sigma_c} = \frac{1.400}{0.140} = 10$$

$$k_d = \sqrt{\frac{2(m+15)^2}{15\sigma_c(m+10)}} = \sqrt{\frac{2(10+15)^2}{15 \cdot 0.140(10+10)}} = 5.46$$

$$M_{T1} = M'_{T1} + C_{1T} \cdot C_T = 18 \cdot 5.11 + 40.0(0.31 - 0.05) = 92.0 + 10.4$$
$$= 102.40 \text{ tm}$$

$$M_i = \frac{b_0 \cdot d^2}{k_d^2} = \frac{40 \cdot 70^2}{5.46^2} = 6575 \text{ tcm} = 65.75 \text{ tm}$$

$$\Delta M'_{T1} = M_{T1} - M_i = 102.40 - 65.75 = 36.65 \text{ tm}$$

$$A_s = \frac{M_i}{z_0 \sigma_s} + \frac{\Delta M'_{T1}}{c \, \sigma_s} - \frac{C_T}{\sigma_s} = \frac{65.75}{0.56 \cdot 1.400} + \frac{36.65}{0.60 \cdot 1.400} - \frac{40.0}{1.400}$$
$$= 83.86 + 43.63 - 28.57 = 98.92 \text{ cm}^2$$

$$\sigma'_s = \frac{15\,\sigma_c}{k_x}\frac{c}{d} - \sigma_s = \frac{15 \cdot 0.140}{0.60}\frac{60}{70} - 1.400 = 3.0 - 1.400 = 1.600 \text{ t/cm}^2$$

$$A'_s = \frac{\Delta M'_{1T}}{c\,\sigma_s} = \frac{36.65}{0.60 \cdot 1.400} = 43.6 \text{ cm}^2$$

Vierendeel—with diagonals (Fig. 4-7):

$$\tan\alpha = \frac{z_0}{c} = \frac{2.40}{3.00} = 0.80 \qquad \alpha = 38°\,40'$$

$$\sin\alpha = 0.625, \qquad \cos\alpha = 0.780$$

$$C_1 = \frac{R}{\tan\alpha} = \frac{18 \cdot 4.00}{0.80} = 90.0 \text{ t}$$

$$D_1 = \frac{C_1}{\cos\alpha} = \frac{90.0}{0.780} \cong 115.0 \text{ t}$$

$$D_2 = \frac{V_2}{\sin\alpha} = \frac{18 \cdot 2.50}{0.625} = 72 \text{ t}$$

$$M'_{T3,\max} = 18 \cdot 2.20 = 39.6 \text{ tm}$$

$$M_{3T} = 39.6 + 152 \cdot 0.26 = 79.12 \text{ tm} \quad M_i = 65.75 \text{ tm}$$

$$\Delta M'_{3T} = 79.12 - 65.75 = 13.37 \text{ tm}$$

$$A_s = \frac{65.75}{0.56 \cdot 1.400} + \frac{13.37}{0.60 \cdot 1.400} - \frac{152.}{1.40} = 83.86 + 5.92 - 108.57 = -$$

$$A'_s = \frac{13.37}{0.60 \cdot 1.400} = 15.92 \text{ cm}^2$$

The moments are illustrated in Fig. 4-9b.

Verticals (critical only):

$$M_{1v} = 157.70 \text{ tm}, \quad P = 18 \text{ t}$$

$$d = k_d\sqrt{\frac{M_{1v}}{b_0}} = 5.46\sqrt{\frac{15770}{40}} = 108 \text{ cm}$$

Use: $d_0 = 100$ cm, $d = 90$ cm $c_t = 42$ cm

$$x = 0.60 \cdot 90 = 54 \text{ cm}, \quad z_0 = 72 \text{ cm}, \quad c = 90 \text{ cm}$$

$$M_i = \frac{b_0 d^2}{k_d^2} = \frac{40 \cdot 90^2}{5.46^2} = 10868 \text{ t cm} = 108.68 \text{ tm}$$

$$\Delta M_1 = 157.70 + 18 \cdot 0.42 - 108.68 = 56.58 \text{ tm}$$

Vierendeel Girders

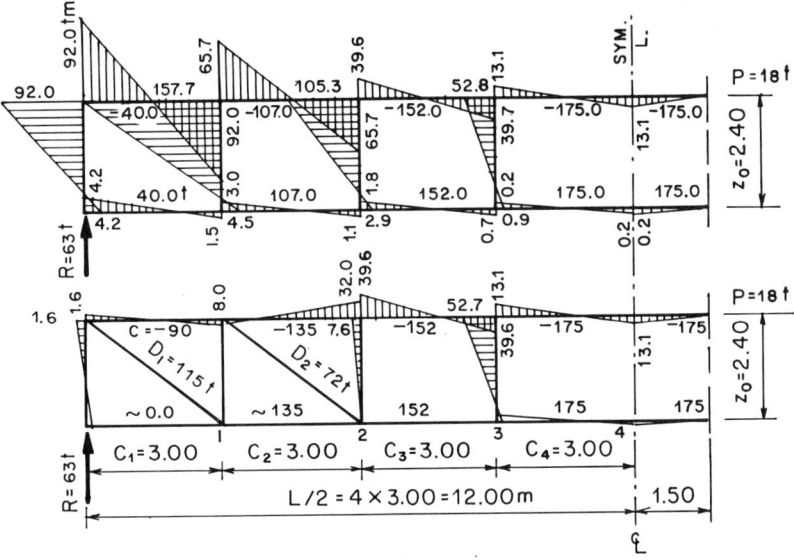

Figure 4-9(a) and (b)

$$A_s = \frac{108.68}{0.72 \cdot 1.400} + \frac{56.58}{0.90 \cdot 1.400} - \frac{18}{1.400} \cong 140 \text{ cm}^2$$

$$A'_s = \frac{56.58}{0.90 \cdot 1.400} = 44.90 \text{ cm}^2, \quad \sigma' = (\sigma_s + 15\sigma_c)\frac{c}{d} - \sigma_s > \sigma_s$$

$$v_{1v} = \frac{67000}{40 \cdot 72} \sim 23 \text{kg/cm}^2 \quad v_{10} = \frac{61300}{40 \cdot 56} = \sim 37 \text{ kg/cm}^2$$

$$D_{1,\max} = 115t, \quad A_{1s} = \frac{115}{1.400} = 82 \text{ cm}^2, \quad A_{2s} = \frac{72}{1.400} = 51 \text{ cm}^2$$

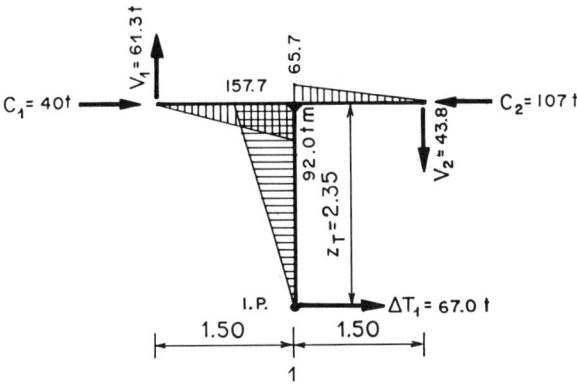

Figure 4-10

Diagonals:
Bottom chord:

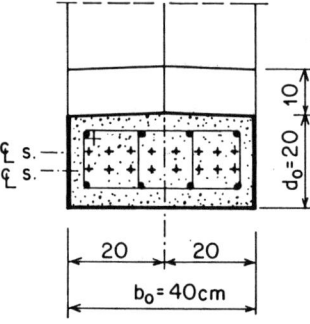

Figure 4-11

$$d_0 = 30 \text{ cm}, \ d = 25 \text{ cm}, \ M'_{1B} = -4.2 \ tm, \ T_1 = 40 \ t$$

$$k_x = 0.60, \ x = 0.60 \cdot 25 = 15 \text{ cm}, \ z_0 = 0.8 \cdot 25 = 20 \text{ cm}$$

$$k_d = 5.46, \ c_T = \frac{30}{2} - d' = 10 \text{ cm}$$

$$M'_{T1} = M'_{B1} - T_1 \cdot c_T = 4.2 - 40 \cdot 0.10 = 0.20 \ tm$$

$$M_i = \frac{40 \cdot 25^2}{5.46^2} = 839 \ t \text{ cm}$$

$$A_s = \frac{M'_{T1}}{z_0 \sigma_s} + \frac{T_1}{\sigma_s} = \frac{0.20}{0.20 \cdot 1.400} + \frac{40}{1.400} = 29.28 \text{ cm}^2$$

$$T_{4,\max} = 175 \ t, \ d_0 = 20 \text{ cm}, \ b'_0 = 30 \text{ cm}$$

$$A_s = \frac{175}{1.400} = 125 \text{ cm}^2 \text{ (use prestressing)}$$

5

Two-way Slabs and Grids

METHOD OF ANALYSIS

For uniform loading (w), the moments in a two-way slab can be determined most simply by Marcus' method.* The method is carried out in two steps. First, the M'_x and M'_y moments are computed by considering the two-way slab as composed of a series of equal-width strips acting independently in the x- as well as y-direction.

The intersecting points of the center lines of each strip, loaded by w_x and w_y, must have the same deflections.

$$\delta'_x = \delta'_y \qquad w_x + w_y = w \qquad (5\text{-}1)$$

The deflection δ-values depend on the loading w_x, w_y, the spans L_x, L_y, and the supporting conditions: simply supported, fixed-end, or a combination of one end simply supported and the other end fixed.

The corresponding midspan deflections δ' for a constant EI are:

---

*H. Marcus, *Vereinfachte Berechnung biegsamer Platten* (Berlin: Springer-Verlag, 1929).

98 Chap. 5 Two-way Slabs and Grids

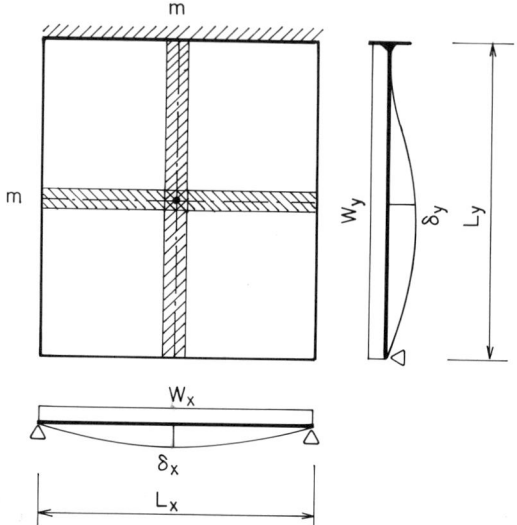

Figure 5-1

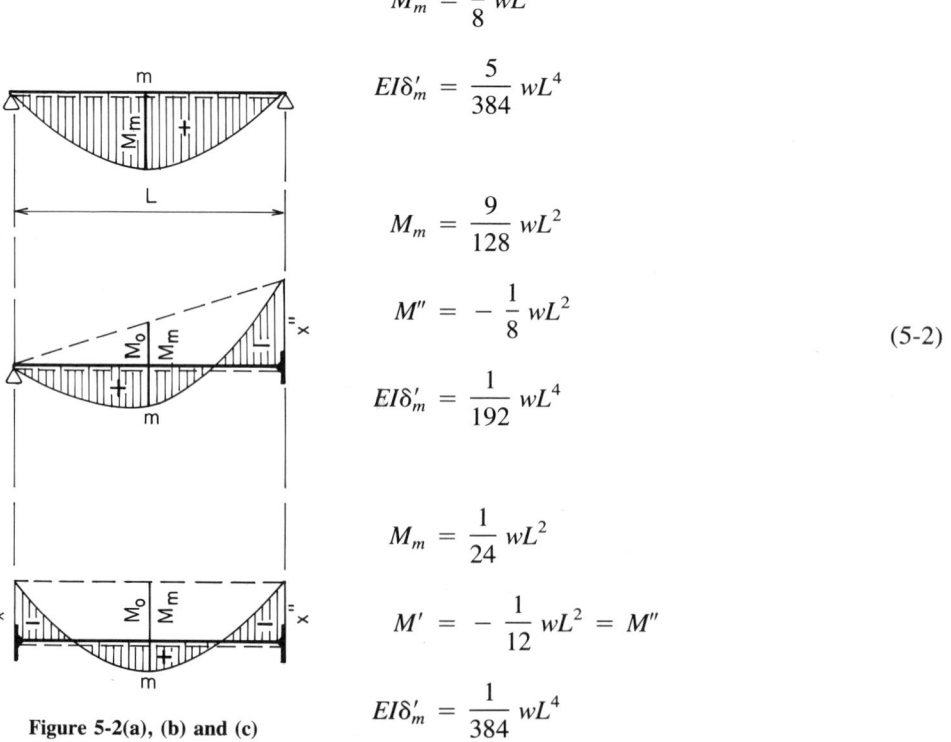

Figure 5-2(a), (b) and (c)

$$M_m = \frac{1}{8} wL^2$$

$$EI\delta'_m = \frac{5}{384} wL^4$$

$$M_m = \frac{9}{128} wL^2$$

$$M'' = -\frac{1}{8} wL^2$$

$$EI\delta'_m = \frac{1}{192} wL^4$$

$$M_m = \frac{1}{24} wL^2$$

$$M' = -\frac{1}{12} wL^2 = M''$$

$$EI\delta'_m = \frac{1}{384} wL^4$$

(5-2)

Method of Analysis

One-Span Slabs

Applying Eq. 5-1 and taking $L_y/L_x = \varepsilon$ and $w_x/w = k_0$ the w_x and w_y relationship for two-way symmetrical boundary conditions (Fig. 5-3a) and the corresponding w_x- and w_y-values become:

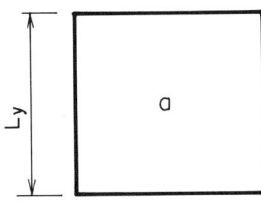

$$w_x = \frac{L_y^4}{L_x^4 + L_y^4} w = k_0 w \qquad k_0 = \frac{\varepsilon^4}{1 + \varepsilon^4}$$

$$w_y = (1 - w_x)w \qquad K_a = k_b = k_c = k_0 \tag{5-3}$$

$$M_x' = \frac{1}{8} k_0 w L_x^2 \qquad M_y' = \frac{1}{8}(1 - k_0) w L_y^2$$

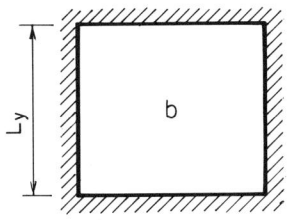

$$M_x' = \frac{1}{24} k_0 \, w L_x^2$$

$$M_y' = \frac{1}{24}(1 - k_0) \, w L_y^2 \tag{5-4}$$

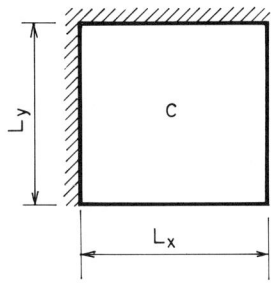

$$M_x' = \frac{9}{128} k_0 \, w L_x^2$$

$$M_y' = \frac{9}{128}(1 - k_0) \, w L_y^2$$

Figure 5-3(a)

For simple symmetry, there are three boundary cases, as given in Fig. 5-3b. The w_x, w_y values and maximum midspan moments are:

$$w_x = \frac{5\varepsilon^4}{2 + 5\varepsilon^4} w = k_1 w \qquad k_1 = \frac{5\varepsilon^4}{2 + 5\varepsilon^4}$$

$$w_y = (1 - k_1)w$$

$$M'_x = \frac{9}{128} k_1 wL_x^2 \qquad M'_y = \frac{1}{8}(1 - k_1)wL_y^2$$

$$w_x = \frac{5\varepsilon^4}{1 + 5\varepsilon^4} w = k_2 w \qquad k_2 = \frac{5\varepsilon^4}{1 + 5\varepsilon^4}$$

$$w_y = (1 - k_2)w \qquad (5\text{-}5)$$

$$M'_x = \frac{1}{24} k_2 wL_x^2 \qquad M'_y = \frac{1}{8}(1 - k_1)wL_y^2$$

$$w_x = \frac{2\varepsilon^4}{1 + 2\varepsilon^4} w = k_3 w \qquad k_3 = \frac{2\varepsilon^4}{1 + 2\varepsilon^4}$$

$$w_y = (1 - k_3)w$$

$$M'_x = \frac{1}{24} k_3 wL_x^2 \qquad M'_y = \frac{9}{128}(1 - k_3)wL_y^2$$

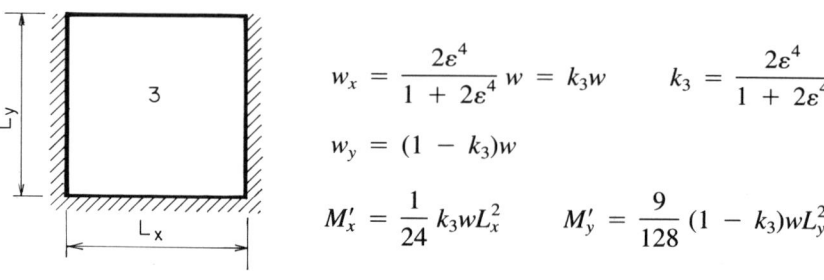

Figure 5-3(b)

Method of Analysis

The continuity of the slab is established by the action of M_{xy} due to the shear between the slab strips. In accordance with Marcus, the influence of the M_{xy} action can be expressed in terms of

$$-M''_x = \varphi_x M'_x \qquad -M''_y = \varphi_y M'_y$$

where

$$\varphi_x = \frac{5}{6}\varepsilon^2 \frac{M'_x}{M_{0x}} \qquad M_{0x} = \frac{1}{8}wL_x^2$$

$$\varphi_y = \frac{5}{6}\varepsilon^2 \frac{M'_y}{M_{0y}} \qquad M_{0y} = \frac{1}{8}wL_y^2$$

(5-6)

For two-way symmetry, simple, fixed and simple fixed-end slabs, the φ-values are, respectively,

$$\text{a:} \quad \varphi_{ax} = \varphi_{ay} = \frac{5}{6}\frac{\varepsilon^2}{1+\varepsilon^4}$$

$$\text{b:} \quad \varphi_{bx} = \varphi_{by} = \frac{5}{18}\frac{\varepsilon^2}{1+\varepsilon^4}$$

(5-7)

$$\text{c:} \quad \varphi_{cx} = \varphi_{cy} = \frac{15}{32}\frac{\varepsilon^2}{1+\varepsilon^4}$$

Simple symmetry:

$$\text{1:} \quad \varphi_{1x} = \frac{75}{32}\frac{\varepsilon^2}{2+5\varepsilon^4} \qquad \varphi_{1y} = \frac{5}{3}\frac{\varepsilon^2}{2+5\varepsilon^4}$$

$$\text{2:} \quad \varphi_{2x} = \frac{25}{18}\frac{\varepsilon^2}{1+5\varepsilon^4} \qquad \varphi_{2y} = \frac{5}{6}\frac{\varepsilon^2}{1+5\varepsilon^4}$$

(5-8)

$$\text{3:} \quad \varphi_{3x} = \frac{5}{9}\frac{\varepsilon^2}{1+2\varepsilon^4} \qquad \varphi_{3y} = \frac{15}{32}\frac{\varepsilon^2}{1+2\varepsilon^4}$$

Thus, the final moments M_x and M_y are:

$$M_x = M'_x(1 - \varphi_x) = c_x w L_x^2$$

$$M_y = M'_y(1 - \varphi_y) = c_y w L_y^2$$

(5-9)

The c-values for all six slab types with varying ε- and k-values as well as moments (M,X) are given in Table 3.

The reactions (R) at the supports are close to a parabola and their total value is roughly

$$\text{Short span:} \quad \Sigma R_x = \frac{1}{4} w L_x^2$$

$$\text{Long span:} \quad \Sigma R_y = \frac{1}{4} w L_x (2L_y - L_x)$$

(5-10)

Example

$$L_y = 7.80 \text{ m} \qquad L_x = 6.00 \text{ m} \qquad w = 0.55 \text{ t/m}^2$$

Slab type b:

$$\varepsilon = \frac{L_y}{L_x} = \frac{7.80}{6.00} = 1.30$$

$$k_0 = \frac{\varepsilon^4}{1 + \varepsilon^4} = \frac{1.30^4}{1 + 1.30^4} = 0.741$$

$$w_x = k_0 w = 0.741 \cdot 0.55 = 0.407 \text{ t/m}^2$$

$$w_y = (1 - k_0) w = 0.143 \text{ t/m}^2$$

$$M'_x = \frac{1}{24} w_x L_x^2 = \frac{1}{24} 0.407 \cdot 6.00^2 = 0.612 \text{ tm}$$

$$M'_y = \frac{1}{24} w_y L_y^2 = \frac{1}{24} 0.143 \cdot 7.80^2 = 0.363 \text{ tm}$$

$$\varphi_x = \varphi_y = \frac{5}{18} \frac{\varepsilon^2}{1 + \varepsilon^4} = \frac{5}{18} \frac{1.30^2}{1 + 1.30^4} = 0.122$$

$$M_x = M'_x (1 - \varphi_x) = 0.612 \cdot 0.878 = 0.537 \text{ tm}$$

$$M_4 = M'_y (1 - \varphi_y) = 0.363 \cdot 0.878 = 0.318 \text{ tm}$$

Slab type 3:

$$\varepsilon = 1.30$$

$$k_3 = \frac{2\varepsilon^4}{1 + 2\varepsilon^4} = \frac{2 \cdot 1.30^4}{1 + 2 \cdot 1.30^4} = 0.851$$

$$w_x = k_3 w = 0.851 \cdot 0.55 = 0.468 \text{ t/m}^2$$

$$w_y = (1 - k_3) w = 0.149 \cdot 0.55 = 0.082 \text{ t/m}^2$$

Method of Analysis

$$M'_x = \frac{1}{24} w_x L_x^2 = \frac{1}{24} \cdot 0.468 \cdot 6.00^2 = 0.702 \text{ tm}$$

$$M'_y = \frac{9}{128} w_y L_y^2 = \frac{9}{128} 0.082 \cdot 7.80^2 = 0.351 \text{ tm}$$

$$\varphi_x = \frac{5}{9} \frac{\varepsilon^2}{1 + 2\varepsilon^4} = \frac{5}{9} \frac{1.30^2}{1 + 2 \cdot 1.30^4} = 0.140$$

$$\varphi_y = \frac{15}{32} \frac{\varepsilon^2}{1 + 2\varepsilon^4} = \frac{15}{32} \frac{1.30^2}{1 + 2 \cdot 1.30^4} = 0.118$$

$$M_x = M'_x (1 - \varphi_x) = 0.702 (1 - 0.140) = 0.604 \text{ tm}$$

$$M_y = M'_y (1 - \varphi_y) = 0.351 (1 - 0.118) = 0.310 \text{ tm}$$

Support moments and span moment coefficients of the slab types a, b, c and 1, 2, 3 for various l_x/l_y ratios are computed and given in Table 3, included in the Appendix.

Continuous Slabs or Grids

For continuous slabs or narrow grid systems (Fig. 5-4) subjected to checkerboard live-load loading conditions, the use of symmetry and antisymmetry allows the maximum span moments to be obtained without any modification of the foregoing method. The method is illustrated in principle in Fig. 5-5.

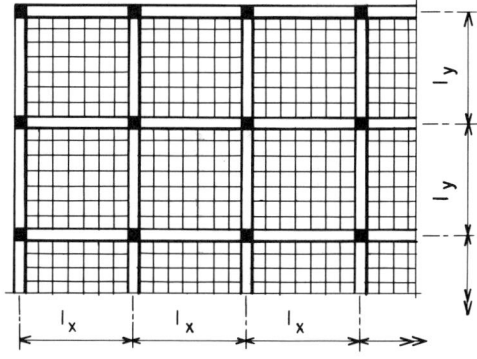

Figure 5-4

For symmetrical parts, all slabs are loaded uniformly with $\frac{1}{2} w_L$ and the moments M_x and M_y are computed as discussed. For antisymmetrical loading, the adjacent slabs are loaded with $\pm \frac{1}{2} w_L$. Under this loading condition, the slabs behave in span direction as independent simply supported slabs. The final results are obtained by superposition ($M = M' + M''$).

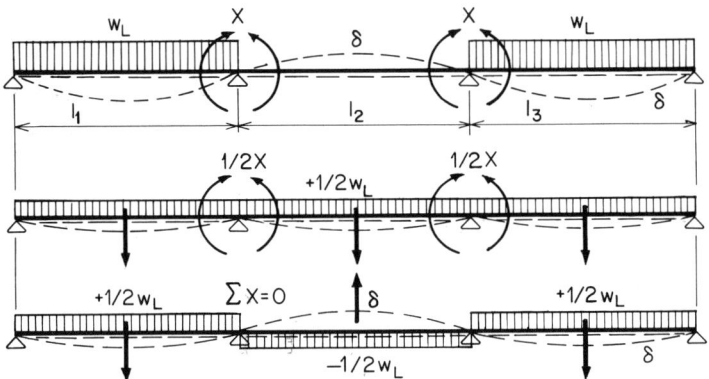

Figure 5-5

A beam grid consists of two sets of beams lying on the same plane and intersecting each other at an angle (Fig. 5-4). The statical behavior of such a beam grid is similar to that of a two-way slab. However, the significance of the M_{xy}-moments in the beam grid is negligible in comparison with the slab. Due to this the stress condition of the beams is adequately described for two-way slab by Marcus' method.

Reinforcing The reinforcing for the slab is commonly determined from the maximum moments M_x, M_y and distributed in equal spacing over the middle strip $L_x/2$. For marginal strips in both directions, the minimum reinforcing is taken $\frac{1}{2} A_s$ of the middle strip reinforcing (Fig. 5-6).

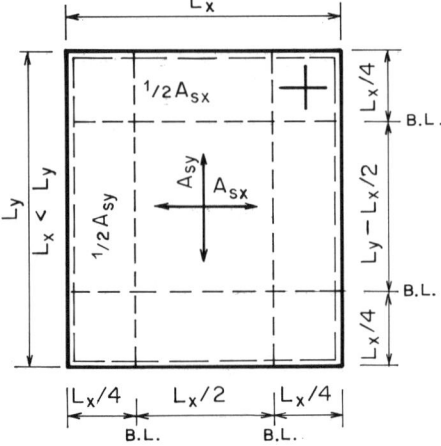

Figure 5-6

Method of Analysis

The free corners (not homogeneously tied to the edge beams) must have the same reinforcing in both directions (top and bottom) as the maximum reinforcing in x-direction of the middle strip.

Slenderness

Simply supported two-way slab:

$$t_{min} = \frac{L_{x0}}{35} L_x < L_y \qquad (5\text{-}11)$$

One end simply supported, the other end fixed:

$$t_{min} = \frac{0.80\, L_{x0}}{35} \cong \frac{L_x}{45}$$

Both ends fixed:

$$t_{min} = \frac{0.60\, L_{x0}}{35} \cong \frac{L_x}{55}$$

Maximum deflection:

$$\delta_{max} \cong \frac{L_x}{360} \qquad (5\text{-}12)$$

Example

For demonstration of the symmetry-antisymmetry method (Fig. 5-5) and the use of Table 3, the maximum midspan and support moments of the continuous two-way slab system, illustrated in Fig. 5-7, will be computed.

$$L_x = 6.00 \text{ m}, \ L_y = 7.80 \text{ m}$$

$$\varepsilon = \frac{L_y}{L_x} = 1.30$$

$$w_D = 0.40 \text{ t/m}^2, \ w_L = 0.30 \text{ t/m}^2$$

$$w = 0.40 + 0.30 = 0.70 \text{ t/m}^2$$

Symmetric loading:

$$w' = w_D + w_L/2 = 0.40 + 0.15 = 0.55 \text{ t/m}^2$$

Antisymmetric loading:

$$w'' = \pm w_L/2 = \pm 0.150 \text{ t/m}^2$$

$$M' = \text{symmetric moment}$$

$$M'' = \text{antisymmetric moment}$$

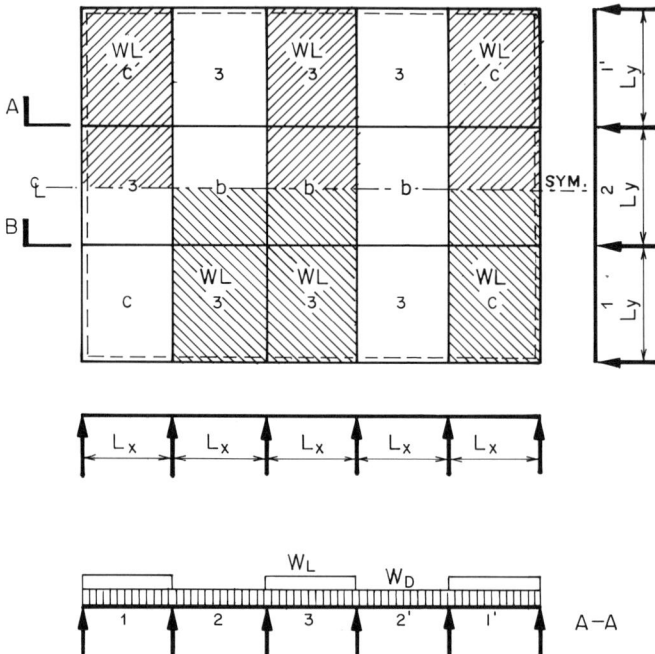

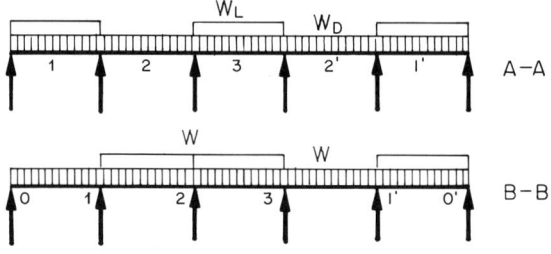

Figure 5-7

Slab Moments (Table 3)

In order to obtain maximum slab moments (M''), a checkerboard pattern of live load is to be used. Making use of symmetry and antimetry, the antimetric moments due to w'' are determined by the use of slab type "a" (simply supported slab). The support moments (x) are computed for uniform load $w = w_0 + w_L$. Thus:

$$M = M' + M'' = (cw' + c_a w'')L^2$$

$$X = -\frac{1}{m} kwL^2, \quad m = 8 \text{ or } 12$$

Span 1
Slab c and a:

$$M_x = (c_{cx}w' + c_{ax}w'')L_x^2$$
$$= (0.041 \cdot 0.55 + 0.059 \cdot 0.15)\, 6.0^2 = 1.13 \text{ tm}$$

Method of Analysis

$$M_y = (c_{cy}w' + c_{ay}w'')L_y^2$$
$$= (0.014 \cdot 0.55 + 0.021 \cdot 0.15)7.8^2 = 0.66 \text{ tm}$$

$$X_{cx} = -\frac{1}{8}k_{cx}wL_x^2 = -\frac{1}{8}0.741 \cdot 0.70 \cdot 6.0^2 = -2.33 \text{ tm}$$

$$X_{cy} = -\frac{1}{8}k_{cy}wL_y^2 = -\frac{1}{8}0.259 \cdot 0.70 \cdot 7.8^2 = -1.38 \text{ tm}$$

Slab 3 and a:

$$M_x = (c_{3x}w' + c_{ax}w'')L_x^2$$
$$= (0.031 \cdot 0.55 + 0.059 \cdot 0.15) 6.0^2 = 0.93 \text{ tm}$$

$$M_y = (c_{3y}w' + c_{ay}w'')L_y^2$$
$$= (0.009 \cdot 0.55 + 0.021 \cdot 0.15) 7.8^2 = 0.49 \text{ tm}$$

$$X_{x3} = -\frac{1}{8}k_{3x}wL_x^2 = -\frac{1}{8}0.149 \cdot 0.70 \cdot 6.0^2 = -0.47 \text{ tm}$$

$$X_{3y} = -\frac{1}{12}k_{3y}wL_y^2 = -\frac{1}{12}0.851 \cdot 0.70 \cdot 7.8^2 = -3.02 \text{ tm}$$

Span 2
Slab 3 and a:

$$M_x = (c_{3x}w' + c_{ax}w'')L_x^2$$
$$= (0.009 \cdot 0.55 + 0.059 \cdot 0.15) 6.0^2 = 0.50 \text{ tm}$$

$$M_y = (c_{3y}w' + c_{ay}w'')L_y^2$$
$$= (0.031 \cdot 0.55 + 0.021 \cdot 0.15) 7.8^2 = 1.23 \text{ tm}$$

$$X_{3x} = -\frac{1}{12}k_{3x}wL_x^2 = -\frac{1}{12}0.851 \cdot 0.70 \cdot 6.0^2 = -1.79 \text{ tm}$$

$$X_{3y} = -\frac{1}{8}k_{3y}wL_y^2 = -\frac{1}{8}0.149 \cdot 0.70 \cdot 7.8^2 = -0.79 \text{ tm}$$

Slab b and a:

$$M_x = (c_{bx}w' + c_{ax}w'')L_x^2$$
$$= (0.027 \cdot 0.55 + 0.059 \cdot 0.15) 6.0^2 = 0.85 \text{ tm}$$

$$M_y = (c_{by}w' + c_{ay}w'')L_y^2$$
$$= (0.009 \cdot 0.55 + 0.021 \cdot 0.15) 7.8^2 = 0.49 \text{ tm}$$

$$X_{bx} = -\frac{1}{12} k_{bx} w L_x^2 = -\frac{1}{12} 0.741 \cdot 0.70 \cdot 6.0^2 = -1.56 \text{ tm}$$

$$X_{by} = -\frac{1}{12} k_{by} w L_y^2 = -\frac{1}{12} 0.259 \cdot 0.70 \cdot 7.8^2 = -0.92 \text{ tm}$$

The slab moments ($M_x X_x$, $M_y X_y$) and the support moment (X_x) distribution in y-direction and (X_y) in x-direction are shown for slab b in Fig. 5-8.

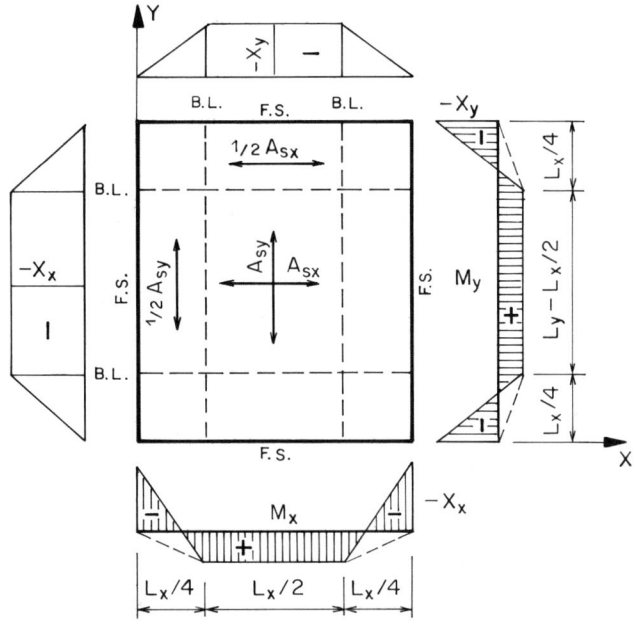

Figure 5-8

As can be seen, the slab moments (X) differ at common marginal boundaries. Such differences induce torsional moments (M_T) into beams. In order to reduce the M_T-moments, the average X-values are used in practice. The influence of ΔX-changes upon the slab moments is commonly neglected mainly because of the load transfer due to creep.

Average slab support moments:

$$X_{1x} = \frac{X_{cx} + X_{3x}}{2} = -\frac{2.33 + 1.79}{2} = -2.06 \text{ tm}$$

$$X_{2x} = \frac{X_{x3} + X_{bx}}{2} = -\frac{0.47 + 1.56}{2} = -1.02 \text{ tm}$$

Method of Analysis

$$X_{1y} = \frac{X_{cy} + X_{3y}}{2} = -\frac{1.38 + 3.02}{2} = -2.20 \text{ tm}$$

$$X_{2y} = \frac{X_{3y} + X_{by}}{2} = -\frac{0.79 + 0.92}{2} = -0.86 \text{ tm}$$

$$X_{3x} = \frac{X_{2x} + X_{3x}}{2} = -\frac{2 \cdot 1.79}{2} = -1.79 \text{ tm}$$

$$X_{bx} = \frac{X_{bx} + X_{bx}}{2} = -\frac{2 \cdot 1.56}{2} = -1.56 \text{ tm}$$

Framing

Commonly the beams in x and y-directions are supported by columns and analyzed as continuous beams or frames. The slab loads transferred from slabs to beams are triangular in x-direction and trapezoidal in y-direction ($L_y > L_x$). This method is rather complicated and inaccurate, resulting in underdesigning of beams, especially in short-span direction. This fact is proven by more accurate analysis.

Adopting the flat-slab method of analysis, the whole structural system is considered as a unit. The beams or frames are analyzed in both directions for full load:

$$\overline{w}_x = wL_y = 0.70 \cdot 7.8 = 5.46 \text{ t/m} \tag{5.13}$$

$$\overline{w}_y = wL_x = 0.70 \cdot 6.0 = 4.20 \text{ t/m}$$

The support moments $\overline{X}_x$, $\overline{X}_y$ and span moments $\overline{M}_x$, $\overline{M}_y$ for this loading are computed by using Table 1. In case the L_x and L_y differ in either direction more than 20 percent, the span-by-span method should be used. The final moments are obtained by deducting the slab moments M_x and M_y computed by the use of Table 3. Thus

$$\Sigma\overline{X}_x = \overline{X}_x - X_x\left(L_y - \frac{L_x}{4}\right)$$

$$\Sigma\overline{X}_y = \overline{X}_y - X_y\left(L_x - \frac{L_x}{4}\right)$$

$$\Sigma\overline{M}_{mx} = \overline{M}_{mx} - M_{mx}\left(L_y - \frac{L_x}{4}\right) \tag{5.14}$$

$$\Sigma\overline{M}_{my} = \overline{M}_{my} - M_{my}\left(L_x - \frac{L_x}{4}\right)$$

For common loading the $\overline{X}$ and $\overline{M}$ moments could be determined for uniform loading on all spans. However, where live load is relatively high, movable live load (Fig. 2-27) should be used.

This method is rather accurate and relatively simple, as will be demonstrated in the following.

Five-span beams in x-direction (Fig. 5-7):

$$\overline{w}_x = wL_y + w_{DB} = 0.70 \cdot 7.8 + 0.06 = 5.52 \text{ t/m}$$

Span 1:

$$\overline{X}_1 = -[(0.0625 + 0.0500 + 0.003) - (0.0125 + 0.001)]\overline{w}_x L_x^2$$

$$= -(0.1156 - 0.0135) \cdot 5.52 \cdot 6.0^2 = -20.29 \text{ tm}$$

$$\Sigma \overline{X}_1 = \overline{X}_1 - \frac{X_{1x} + X_{2x}}{2}\left(L_y - \frac{L_x}{4}\right) = -20.29 + \frac{2.06 + 1.02}{2} 6.3$$

$$= -10.59 \text{ tm}$$

$$\overline{M}_{1m} = [(0.0950 + 0.005 + 0.0004) - (0.0200 + 0.0013)]\overline{w}_x L_x^2$$

$$= (0.1004 - 0.0213) 5.52 \cdot 6.0^2 = 15.72 \text{ tm}$$

$$\Sigma \overline{M}_{1m} = \overline{M}_{1m} - \frac{M_{cx} + M_{3x}}{2}\left(L_y - \frac{L_x}{4}\right) = 15.72 - \frac{1.130 + 0.93}{2} 6.3$$

$$= 9.23 \text{ tm}$$

Span 2:

$$\overline{X}_2 = -[(2 \cdot 0.0500 + 0.0039) - (0.0156 + 0.0125)]\overline{w}_x L_x^2$$

$$-(0.1039 - 0.0281) 5.52 \cdot 6.0^2 = -15.06 \text{ tm}$$

$$\Sigma \overline{X}_2 = \overline{X}_2 + \frac{X_{3x} + X_{bx}}{2}\left(L_y - \frac{L_x}{4}\right) = -15.06 + \frac{1.79 + 1.56}{2} 6.3$$

$$= -4.51 \text{ tm}$$

$$\overline{M}_{2x} = [(0.075 + 0.0047) - (0.0234 + 0.0188 + 0.0015)]\overline{w}_x L_x^2$$

$$= (0.0797 - 0.0437) 5.52 \cdot 6.0^2 = 7.15 \text{ tm}$$

$$\Sigma \overline{M}_{2x} = \overline{M}_{2x} - \frac{M_{cx} + M_{bx}}{2}\left(L_y - \frac{L_x}{4}\right) = 7.15 - \frac{0.500 + 0.85}{2} 6.3$$

$$= 2.90 \text{ tm}$$

Reactions (Eq. 5-10):

$$R_{0x} = \frac{2}{4 \cdot 2} wL_x^2 = \frac{1}{4} 0.70 \cdot 6.0^2 = 6.30 \text{ t}$$

$$R'_{1x} = R_{0x} - \frac{\Sigma \overline{X}_1}{L_x} = 6.30 - \frac{10.59}{6.0} = 4.54 \text{ t}$$

Method of Analysis

$$R''_{1x} = R_{0x} + \frac{\Sigma \overline{X}_1}{L_x} = 6.30 + 1.76 = 8.06 \text{ t}$$

$$R_{2x} = 2R_{0x} = 2 \cdot 6.30 = 12.60 \text{ t}$$

Beams in y-Direction

The analysis of beams in y-direction is the same as for x-direction. In order to eliminate the "column forest," the slab reactions in y-direction and beam reactions in x-direction are carried by a simple or continuous frame from wall to wall.

Example

Simple prestressed frame:
Span:
$$L = 3 \cdot L_y - d_0 = 3 \cdot 7.8 - 1.0 = 22.4 \text{ m}$$

Loading (Eq. 5-10):

$$\Sigma R_y = \frac{2}{4} w L_x (2L_y - L_x) = \frac{1}{2} \cdot 0.70 \cdot 6.0 (2 \cdot 7.8 - 6.0) = 20.16 \text{ t}$$

$$\overline{w} = \frac{\Sigma R_y}{L_y} + w_{D,B} = \frac{20.16}{7.8} + 0.60 = 3.18 \text{ t/m}$$

$$P = R_{2x} + R_B = 2 \cdot 6.3 + 0.06 \cdot 6.0 = 12.96 \text{ t}$$

Setting in Fig. 3-23 $z_e = 0$ we obtain:

$$\lambda = \frac{L}{h} = \frac{22.4}{5.0} = 4.480$$

$$\kappa = \frac{h}{L} \frac{I_B}{I_h} = \frac{5.0}{22.4} \frac{1.0}{1.0} = 0.223$$

$$\mu = 3 + 2\kappa = 3 + 2 \cdot 0.223 = 3.446$$

$$\xi = \frac{L_y - \frac{1}{2} d_0}{L} = \frac{7.8 - 0.5}{22.4} \cong 0.333$$

$$\xi' = (1 - \xi) = 0.670$$

Thrusts:

$$H_{\overline{w}} = \frac{\lambda}{4\mu} \overline{w} L = \frac{4.480}{4 \cdot 3.446} \cdot 3.18 \cdot 22.4 = 23.15 \text{ t}$$

$$H_p = \frac{3\lambda}{2\mu} 2P\xi\xi' = \frac{3 \cdot 4.480}{2 \cdot 3.446} 2 \cdot 12.96 \cdot 0.333 \cdot 0.670 = 11.27 \text{ t}$$

$$\Sigma H = H_{\overline{w}} + H_p = 23.15 + 11.27 = 34.42 \text{ t}$$

Moments:

$$M_{0m} = \frac{1}{8}\overline{w}L^2 + P\left(\frac{L}{2} - \frac{L_y}{2}\right) = \frac{1}{8} 3.18 \cdot 22.4^2 + 12.96 (11.2 - 3.9)$$

$$= 199.45 + 94.61 = 294.06 \text{ tm}$$

$$M_m = M_{0m} - \Sigma H \cdot h = 294.06 - 34.42 \cdot 5.00 = 121.96 \text{ tm}$$

$$M_{d,e} = \Sigma H \cdot h = -172.10 \text{ tm}$$

Reactions:

$$R = \overline{w}\frac{L}{2} + P = 3.18 \cdot \frac{22.4}{2} + 12.96 = 48.58 \text{ t}$$

FLAT SLABS

Flat slabs are two-way reinforced slabs supported directly by columns. The connections of the slabs and columns are considered rigid. For heavily loaded flat slabs with variable live loads, the capitals of the columns in most cases are required to keep the shear stresses within safe limits. The column types commonly used are shown in Fig. 5-9.

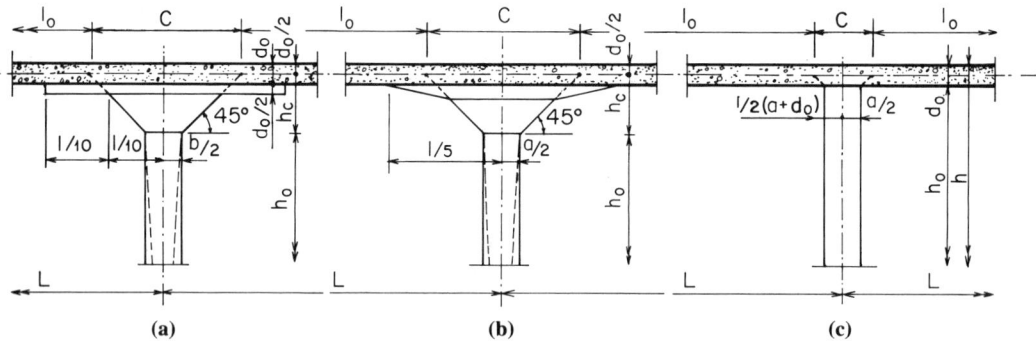

Figure 5-9

The portion of the column capital that lies outside the 45° angle is considered to be nonexistent in structural analysis. The columns can be rectangular, square, round, straight, or tapered. The thickness of the slab must not be less than 1/36 of the largest free span (L_{min}) and not less than 15 cm.

Elastic analysis in accordance with slab theory, numerous tests, and experience prove that the structural behavior of flat-slab structures is rather close to that of multistory and multibay frames. Thus, for structural analysis of flat slabs, the structure is considered to be divided into frames, each consisting of a row of columns and the strips of slabs bounded laterally by the centerline of the panel on either side of the centerline of columns. The division and analysis are carried out longitudinally and transversally for full load ($w = w_D + w_L$) (Fig. 5-10).

Flat Slabs

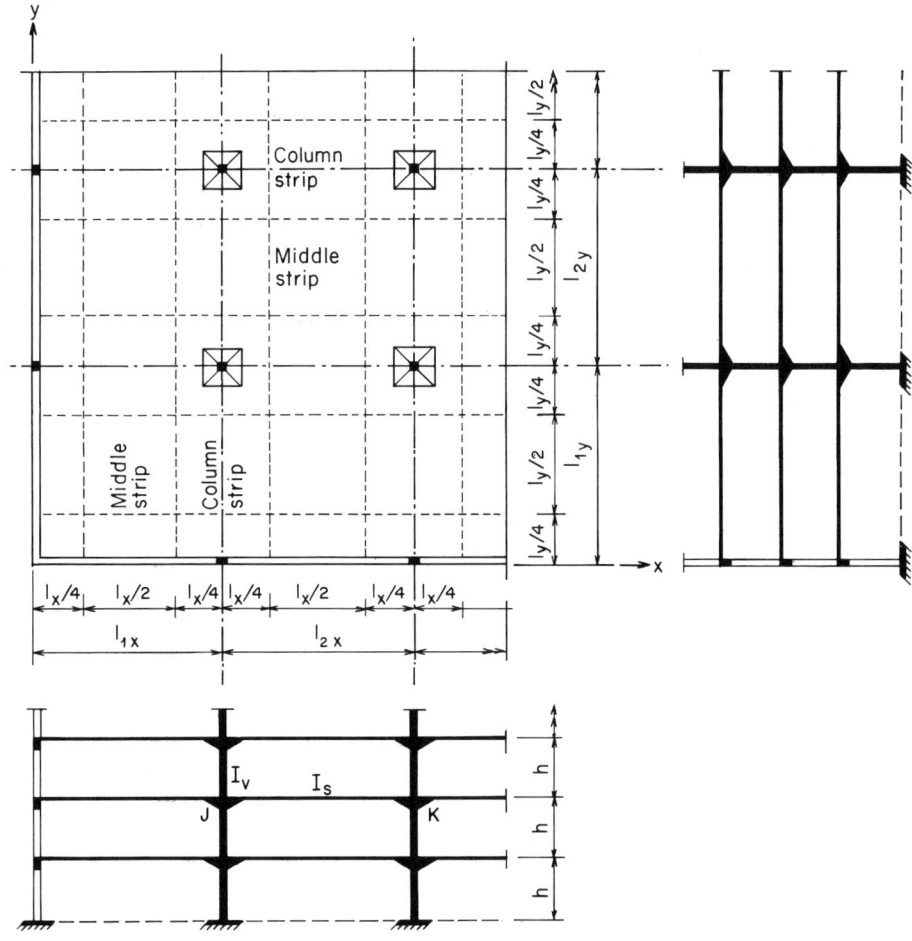

Figure 5-10

The critical moments computed in accordance with frame theory must be divided between column strip c and middle strip m as follows:

$$\text{Column strip } (b_c = 0.5L): M_m^{(c)} = 0.55\, M_m \quad M_x^{(c)} = 0.75\, M_x \quad (5\text{-}15)$$

$$\text{Middle strip } (b_m = 0.5L): M_m^{(m)} = 0.45\, M_m \quad M_x^{(m)} = 0.25\, M_x$$

For flat slabs having no capitals, in accordance with slab theory, the width of the column strips should be $0.40\, L_x$ or $0.40\, L_y$ and the width of the middle strip should be $0.60\, L_x$ or $0.60\, L_y$, whereas half of the total reinforcing of the column strip should be placed on a width $b = a + 2\, d_0$.

Since the slabs are relatively flexible, two adjacent floors above and below and three adjacent bays are considered adequate to compute critical moments. In such an analysis, the columns are assumed fixed or joined at their remote ends. However, for wind loads (nonsymmetric loads), temperature changes, and shrinkage, the entire structure has to be considered.

TWO-SKIN AND WAFFLE SLABS

To reduce dead load, a narrowly spaced grid system can be used instead of a solid slab. Such a system can be structurally analyzed as a flat slab provided that the flexural rigidity of column and middle strips is equal. If this is not the case, the load distribution, as given by Eq. 5-13, is not valid and should be modified in accordance with the rigidity of the strips (Fig. 5-11).

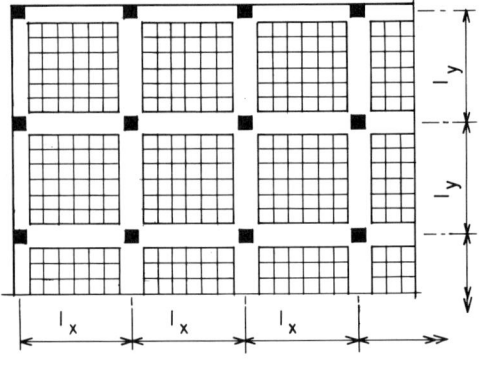

Figure 5-11

Also, for longer spans it is often more economical to use more rigid, prestressed column strips. By prestressing, the deflection can be reduced to zero and the waffle or two-skin grid slab between the column strips can be transferred to a true two-way slab action.

The capital influence can best be taken in account by reducing the span length L_x and L_y by $c/2$ (Fig. 5-9).

Since the slab and columns act structurally together as a rigid frame, the column dimensions should not be less than $l/20$ and the slenderness ratio should not be less than $h/15$ or 30 cm.

The columns at the outside boundary of the flat slab are subjected to rather large moments due to unsymmetric loadings, shrinkage, and temperature changes, and therefore should be rather heavy. To reduce the column sizes and undue deflections of the marginal slabs, edge beams can be used. The connections between the beams and columns should be designed to be flexible to reduce the torsional stresses in beam and the bending stresses in columns.

For longer spans where the slab thickness required is more than 20 cm, use of a solid slab becomes rather uneconomical. Therefore, a two-skin slab with a rigid filler between the slabs (Fig. 5-12b) or a waffle slab (Fig. 5-12a) are far more economical.

Two-skin and Waffle Slabs

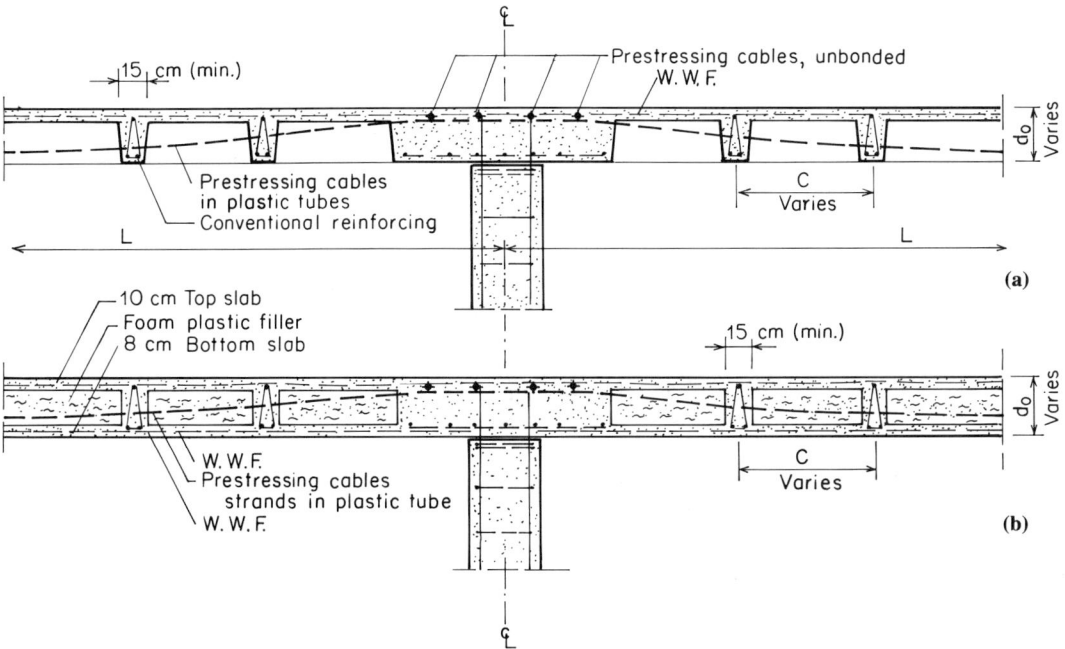

Figure 5-12(a) and (b)

Maximum slendernesses:

Simply supported slab:

$$d_{0,\min} = \frac{L}{35} > 20 \text{ cm}$$

One end simply supported and other elastically fixed:

$$d_{0,\min} = \frac{L}{45} > 20 \text{ cm}$$

Both ends elastically fixed:

$$d_{0,\min} = \frac{L}{55} > 20 \text{ cm}$$

To secure proper transfer of stresses from slab into columns, a solid strip over the column in both directions should be used (Fig. 5-12a,b).

6

Dimensioning and Ultimate Carrying Capacity

GENERAL

In dimensioning and stress analysis, two criteria must be considered: allowable stresses and acceptable deflections. The first criterion determines the dimensions and the second determines the slenderness of a structural element under consideration. Beyond a certain slenderness, excessive deflections and vibration may occur that must be avoided. Both these criteria are related to Eqs. 1-1 and 1-2 (Chapter 1).

The modulus of elasticity E_c is not constant because concrete is not a solid material. It contains a large number of voids and capillary tubes filled with water, moisture, and air, and as such it is subject to capillary laws and stresses. This lends to concrete physical characteristics that change with time, relative humidity, stresses due to loading, increased hydration, and temperature change. Thus, E_c-value can be determined only for time $t = 0$ when plastic deformations are zero. In practice, the E_0-value is estimated with sufficient accuracy as a function of ultimate strength (σ_{uL}) by ROS formula (Eq. 1-13)

$$E_0 = 550000 \frac{\sigma_{uL}}{150 + \sigma_{uL}}$$

where σ_{uL} is the 28-day cylinder stress.

General

For example, $\sigma_{uL} = 300$ kg/cm^2

$$E_0 = 550000 \frac{300}{150 + 300} = 370000 \text{ kg/cm}^2$$

and, respectively, for $\sigma_{uL} = 400$ kg/cm^2,

$$E_0 = 405000 \text{ kg/cm}^2$$

For light-weight concrete, the E_0-value depends on aggregate, which varies considerably. Therefore, the E_t-value has to be determined by tests.

The modulus of elasticity for steel is practically constant

$$E_s = 2100000 \text{ kg/cm}^2$$

Since in reinforced concrete steel and concrete both participate in carrying action, in practice their ratio $n = 15$ is commonly used.

Over the course of time, the E_0-value decreases due to plastic deformations of concrete. This phenomenon is taken into account by the use of $n = 15$ for dimensioning and stress analysis. Thus, the apparent E_t-value for $t = n$:

$$E_t = \frac{E_s}{n} = \frac{2100000}{15} = 140000 \text{ kg/cm}^2 \qquad (6\text{-}1)$$

This means that the steel stress (σ_s) is fifteen times that of the concrete stress (σ_c) and sets the limit for steel participation.

However, for instant- and short-duration loading, the E_0-value is used in stress analysis.

The shear strength of concrete has two features: flexural and direct strength. In accordance with Mohr's theory, the flexural strength is

$$V_F = \frac{1}{2}\sqrt{\sigma_{uL} \cdot \sigma_T} \qquad (6\text{-}2)$$

and the direct strength is

$$V_d = \sqrt{\sigma_{uL} \cdot \sigma_T}$$

The bonding stress (u) depends on the type of reinforcing bars and the quality of the concrete. This is most often determined by tests. For high-quality concrete, $u \cong 8$ kg/cm^2 is used.

In general, the factor of safety (F.S.) for concrete is recommended by code

$$\text{F.S.} = \frac{\sigma_{uL}}{\sigma_c} = 2.5 \qquad \sigma_c = 0.4\sigma_{uL}$$

$$\text{F.S.} = \frac{\sigma_{uL}}{\sigma_c} \cong 1.7 \qquad \sigma_c = 0.6\sigma_{uL}\text{—temporarily}$$

For steel, the allowable stresses are controlled by the acceptable width of cracks—0.3 mm in protected environments and 0.1 to 0.2 mm in severe climates. Due to these limitations, steel is not effectively used in reinforced concrete. This is especially true for high-strength steel grades A432 and A431.

Allowable Stresses

Concrete:

$$\sigma_{uL} = 300 \text{ kg/cm}^2 - \sigma_c = 120 \text{ kg/cm}^2$$

$$\sigma_c = 180 \text{ kg/cm}^2$$

$$\sigma_{uL} = 350 \text{ kg/cm}^2 - \sigma_c = 140 \text{ kg/cm}^2$$

$$\sigma_c = 210 \text{ kg/cm}^2$$

$$\sigma_{uL} = 400 \text{ kg/cm}^2 - \sigma_c = 160 \text{ kg/cm}^2$$

$$\sigma_c = 240 \text{ kg/cm}^2$$

$$\sigma_T \cong \frac{1}{10} \sigma_{uL} \text{—tensile stress}$$

$$v_{uL} = \frac{1}{2}\sqrt{\sigma_{uL} \cdot \sigma_T} = \frac{1}{2}\sqrt{300 \cdot 30} = 47 \text{ kg/cm}^2$$

$$v_o = \frac{v_{uL}}{\text{F.S.}} \sim 20 \text{ k/cm}^2$$

$$u \cong 8 \text{ kg/cm}^2$$

Reinforcing steel:

$$A15: \sigma_y = 2.80 \text{ t/cm}^2$$

$$A432: \sigma_y = 4.22 \text{ t/cm}^2$$

$$A431: \sigma_y = 5.27 \text{ t/cm}^2$$

Prestressing strands:

$$\sigma_{uL} = 19.0 \text{ t/cm}^2\text{—stress released}$$

$$\sigma_{pr} = 0.75\, \sigma_{uL} = 14.25 \text{ t/cm}^2$$

$$\sigma_{pr} = 0.80\, \sigma_{uL} = 15.20 \text{ t/cm}^2\text{—temporarily}$$

Recommended Slenderness Limits

Simple spans:

$$d_o = \frac{L}{16} \quad (d_o\text{—depth of beam, } L\text{—span})$$

One end fixed:

$$d_o = \frac{L - a''}{16} = \frac{0.8\,L}{16} = \frac{L}{20}$$

Both ends fixed and continuous beams:

$$d_o = \frac{L - 2a}{16} = \frac{0.6\,L}{16} \cong \frac{L}{25}$$

For roofs slenderness may be increased about 15%.
  Girders:

$$d_o = \frac{L}{10} \text{ to } \frac{L}{16}$$

SIMPLE BENDING

The theory of simple bending is based on cracked section (assumption 3) and on $n = 15$ for all types of concrete strength. The distance x of the neutral plane (N.P.) from the compressive fiber of the section is obtained from the linear stress distribution over the cross-sectional area (assumption 1). See Fig. 6-1.

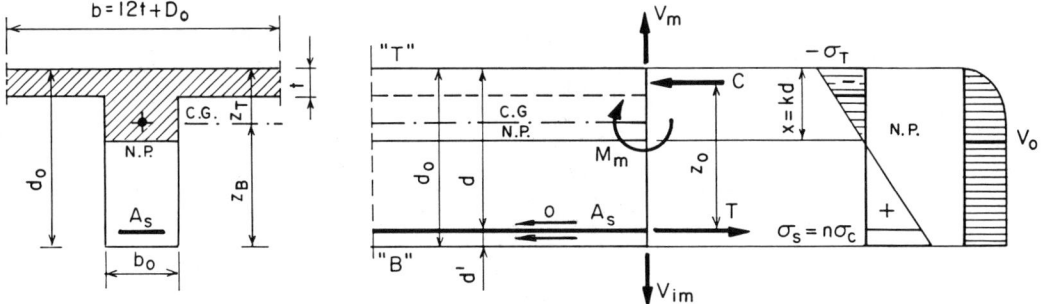

Figure 6-1

$$\frac{\sigma_c + \sigma_s/n}{d} = \frac{\sigma_c}{x} \qquad n = 15$$

$$x = \frac{\sigma_c}{\sigma_c + \sigma_s/n}\,d = \frac{n\sigma_c}{n\sigma_c + \sigma_s}\,d = k_x d \qquad (6\text{-}3)$$

$$k_x = \frac{15\sigma_c}{15\sigma_c + \sigma_s} = \frac{15}{15 + m} \qquad m = \sigma_s/\sigma_c$$

The lever arm (z_0) of the resultants (C, T) of internal stresses is

$$z_0 = \left(1 - \frac{x}{3}\right)d = \frac{10 + m}{15 + m}\,d \qquad (6\text{-}4)$$

In architectural engineering, the compressive stresses below the slab are often neglected ($b \to b_0$). Thus, the internal forces and stresses in accordance with Eq. 1-5 are

$$T = -C = \frac{M_m}{z_0}$$

$$-C = A \quad \sigma_c = \frac{xb_0}{2} \sigma_c \quad \sigma_c = \frac{2T}{xb_0} \qquad (6\text{-}5)$$

$$T = A_s \sigma_s \quad A_s = \frac{T}{\sigma_s} = \frac{M_m}{z_0 \sigma_s}$$

Allowable internal moment (M_i) for chosen concrete stress (σ_c) and the beam depth (d) must be equal to external moment ($M_m X$); thus,

$$M_i = C_c z_0 = \frac{1}{2} \sigma_c x b_0 z_0$$

$$= \frac{1}{2} \sigma_c k_x d \left(1 - \frac{k_x}{3}\right) b_0 d \qquad (6\text{-}6)$$

$$= \frac{1}{2} k_x \left(1 - \frac{k_x}{3}\right) \sigma_c b_0 d^2 = M_m$$

To satisfy the Eq. 6-6, the required depth of the section considered becomes

$$d = \sqrt{\frac{2(m+15)^2}{15\sigma_c (m+10)} \frac{M_m}{b_0}} = k_d \sqrt{\frac{M_m}{b_0}}$$

$$k_d = \sqrt{\frac{2(m+15)^2}{15\sigma_c(m+10)}} = \sqrt{\frac{6}{\sigma_c(3k_x - k_x^2)}} \qquad (6\text{-}7)$$

$$m = \sigma_s/\sigma_c$$

and the internal moment (M_i) (by Eq. 6-7)

$$M_i = \frac{b_0 d^2}{k_d^2} \qquad (6\text{-}8)$$

If the internal moment (M_i) is smaller than the external moment ($M_m X$),

$$\Delta M = M_m - M_i$$

compressive reinforcing is required to balance ΔM. The steel area (A_s, A_s') is determined from Eq. 6-9

Simple Bending

$$\Sigma A_s = \frac{M_i}{z_0 \sigma_s} + \frac{\Delta M}{c} \frac{1}{\sigma_s} \qquad (c = d - 2d') \qquad (6\text{-}9)$$

$$A'_s = \frac{\Delta M}{c \sigma_s} \frac{d - x}{x - d'}$$

Shear stresses (v_0) and bond stress (u) in accordance with Eq. 1-5 are

$$v_0 = \frac{V}{z_0 b_0} \ 1000 \ \text{kg/cm}^2 \qquad (6\text{-}10)$$

$$u = \frac{V}{z_0 \Sigma o} \ 1000 \ \text{kg/cm}^2$$

The stirrups' area (A_v) required to balance the shear stresses (v_0) and spacing (s) for U-stirrups are:

$$A_v = \frac{v_0 b_0}{\sigma_s} \ 100 \ cm^2/m \qquad s = \frac{A_v}{2 a_v} \qquad (6\text{-}11)$$

Example

A three-span continuous beam with span moment $M_{m1} = 14.00$ tm, support moment $X_1 = -18.00$ tm and maximum shear $V''_1 = 10.65$ t, $V'_1 = 7.05$ t will be analyzed in the following.

Support section : (Fig. 6-2)

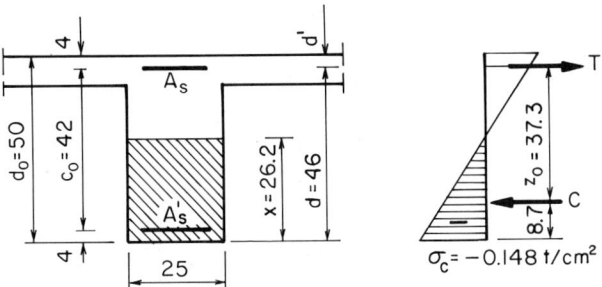

Figure 6-2

$$\sigma_c = 0.120 \ t/cm^2, \ \sigma_s = 1.400 \ t/cm^2$$

$$k_x = \frac{15 \cdot 0.120}{15 \cdot 0.120 + 1.400} = 0.563 \qquad d = 46 \ cm$$

$$x = k_x d = 0.563 \cdot 46 = 26.3 \ cm$$

$$z_0 = \left(1 - \frac{k_x}{3}\right) d = \left(1 - \frac{0.563}{3}\right) 46 = 37.5 \ cm$$

Chap. 6   Dimensioning and Ultimate Carrying Capacity

$$-C_c = T = \frac{X_1}{z_0} = \frac{1800}{37.5} = 48 \text{ t}$$

$$\sigma_c = \frac{2T}{xb_0} = \frac{2 \cdot 4800}{26.3 \cdot 25} = 0.146 > 0.120 \text{ kg/cm}^2$$

$$m = \sigma_s/\sigma_c = 1.400/0.120 = 11.67$$

$$k_d = \sqrt{\frac{2(m+15)^2}{15\sigma_c(m+10)}} = \sqrt{\frac{2(11.67+15)^2}{15 \cdot 0.120 \,(11.67+10)}} = 5.44$$

$$M_i = \frac{b_0 d^2}{k_d^2} = \frac{25 \cdot 46^2}{5.44^2} = 1787 \text{ t cm}$$

$$\Delta M = X - M_i = 1800 - 1787 = 13 \text{ t cm}$$

$$\Sigma A_s = \frac{M_i}{z_0 \sigma_s} + \frac{\Delta M}{c_0 \sigma_s} = \frac{1787}{37.5 \cdot 1.400} + \frac{13}{42 \cdot 1.400}$$

$$= 34.04 + 0.22 = 34.26 \text{ cm}^2$$

$$A'_s = \frac{\Delta M}{c_0 \sigma_s} \frac{d-x}{x-d'} = \frac{13}{42 \cdot 1.400} \frac{46 - 26.3}{26.3 - 4} = 0.20 \text{ cm}^2$$

$$v''_1 = \frac{V''_1}{z_0 b_0} = \frac{10650}{37.5 \cdot 25} = 11.36 \text{ kg/cm}^2$$

$$A_v = \frac{v''_1 b_0}{\sigma_s} \cdot 100 = \frac{11.36 \cdot 25}{1400} \, 100 = 20.3 \text{ cm}^2/\text{m}$$

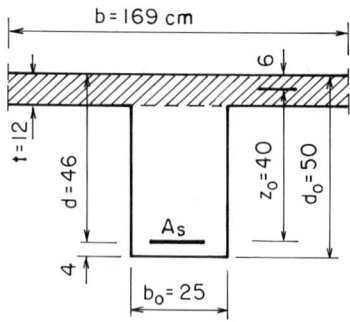

Figure 6-3

$$M_{m1} = 1400 \text{ t cm}$$

$$V'_1 = 7.05 \text{ t}$$

$$b_{\max} = 12 \text{ t} + b_0 = 12 \cdot 12 + 25 = 169 \text{ cm}$$

$$b_{\min} = 6 \text{ t} + b_0 = 6 \cdot 12 + 25 = 97 \text{ cm}$$

Combined Bending and Axial Normal Force

$$z_0 = d - \frac{t}{2} = 46 - \frac{12}{2} = 40 \text{ cm}$$

$$-C = T = \frac{M_{m1}}{z_0} = \frac{1400}{40} = 35.0 \text{ t}$$

$$A_s = \frac{T}{\sigma_s} = \frac{35.0}{1.400} = 25 \text{ cm}^2$$

$$v_1' = \frac{V_1'}{z_0 \cdot b_0} = \frac{7050}{40 \cdot 25} = 7.05 \text{ kg/cm}^2$$

$$\sigma_{c,min} = \frac{C}{b_{max} \cdot t} \sim \frac{35000}{169 \cdot 12} \cong 17 \text{ kg/cm}^2$$

$$\sigma_{c,max} = \frac{35000}{47 \cdot 12} \cong 30 \text{ kg/cm}^2$$

$$A_v = \frac{v_1' b_0}{\sigma_s} 100 = \frac{7.05 \cdot 25}{1400} 100 = 12.6 \text{ cm}^2/\text{m}$$

As can be seen from this example, for a T-section subject to bending the compressive stresses in slab are very small. Therefore, the compressive stresses in stem are often neglected. As a result, the lever arm (z_0) for the T-section $z_0 = d - t/2$ is fully justified.

Further, it seems unrealistic for high-strength concrete to use $n = 15$, resulting in a drop of E_t to $\frac{1}{3} E_0$. However, practical experience and long-lasting tests prove this to be reasonable. This is explained by the formation of hairline cracks in addition to plastic deformation due to shrinkage and creep of concrete.

COMBINED BENDING AND AXIAL NORMAL FORCE

For dimensioning of reinforced concrete members subjected to external moment ($M_m X$), shear (V), and normal force (N) acting at the center of gravity, two cases exist. First, N is rather small force and the eccentricity (e_m) is large outside the kern $\left(\frac{d_0}{3}\right)$

$$e_m = \frac{M}{N} > \frac{d_0}{2 \cdot 3}$$

resulting in large tensile stresses in the section. Second, N is relatively large, thus

$$e_m = \frac{M}{N} \leq \frac{d_0}{6}$$

resulting in no tensile stresses in the section.

However, if the tensile stresses are no more than 1/4 of σ_c-compressive stresses, the second case is still applicable.

For both cases, the three equations of equilibrium (Eq. 1-4) must be satisfied:

$$C_c + C_s - \Sigma T = N$$

$$Ne_T - C_c z_0 - C_s c_0 = 0 \qquad e_T = e_m + c_T \qquad (6\text{-}12)$$

$$V_i = V_m$$

where Ne_T is the external moment and $C_c z_0$ and $C_s c_0$ are the internal moments about the center of reinforcing steel (Fig. 6-4).

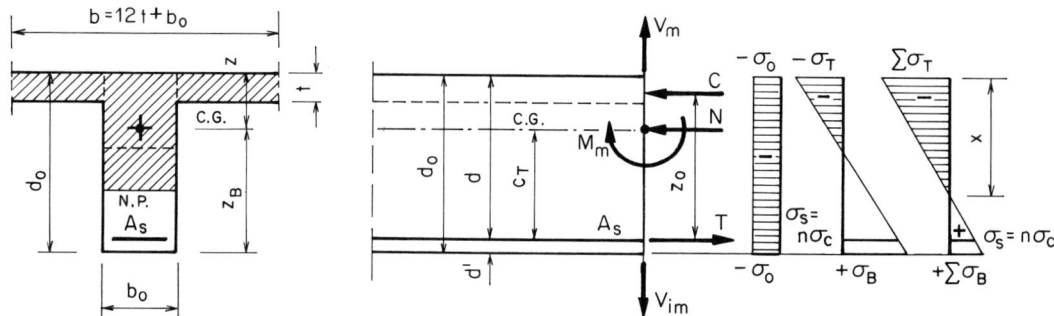

Figure 6-4

Large Eccentricity

Stress analysis of the combined bending can be considered to be simple bending due to external moment M,X superimposed by center stresses due to the normal force N.

$$e_m = \frac{X}{N} > \frac{d_0}{6}$$

We express the C- and T-values (Fig. 6-4) by

$$-C_c = T = A_s \sigma_s \qquad (N = 0)$$

$$-C_s = A'_s n \sigma_c \, c_0 d \qquad (6\text{-}13)$$

$$-\Sigma C_c = C_c + C_s$$

$$\Sigma T = (A_s + A'_s)\sigma_s$$

Introducing these values into Eq. 6-12 and dividing by σ_s, we obtain the total tensional reinforcing area, (ΣA_s).

$$A_s + A'_s - \frac{N}{\sigma_s} = \Sigma C_c = \frac{M_i}{z_0 \sigma_s} + \frac{\Delta M}{c_0 \sigma_s} \qquad (6\text{-}14)$$

$$M_T = X + Nc_T = N(e_m + c_T)$$

Combined Bending and Axial Normal Force

$$\Delta M = M_T - M_i$$

$$M_i = \frac{b_0 d^2}{k_d^2}$$

In accordance with Eq. 6-14, the tensile reinforcement (ΣA_s) of support sections subjected to bending moment (X) and normal force (N) can be determined as for simple bending and then reduced by the amount N/σ_s.

The lever arm (Σz_0) of the combined forces is determined by Eq. 6-15.

$$\Sigma T = -\Sigma C_c \qquad (6\text{-}15)$$

$$\Sigma z_0 = \frac{M_T}{\Sigma A_s \sigma_s + N}$$

Example

Using the same section (Fig. 6-2) loaded as for simple bending but additionally subjected to normal force $N = 27$ t, we obtain the following reinforcement.

Support section:

Assumed allowable stresses

$$\sigma_c = 0.140 \text{ t/cm}^2$$

$$\sigma_s = 1.400 \text{ t/cm}^2$$

$$k_x = \frac{15\sigma_c}{15\sigma_c + \sigma_s} = \frac{15 \cdot 0.140}{15 \cdot 0.140 + 1.400} = 0.60$$

$$x = k_x d = 0.60\,(50 - 5) = 27 \text{ cm}$$

$$z_0 = \left(1 - \frac{k_x}{3}\right) d = \left(1 - \frac{0.60}{3}\right) 45 = 36 \text{ cm}$$

$$k_d = \sqrt{\frac{6}{k_x(3 - k_x)\sigma_c}} = \sqrt{\frac{6}{0.60(3 - 0.60)\,0.140}} = 5.46$$

$$k_d^2 = 29.76$$

$$M_i = \frac{b_0 d^2}{k_d^2} = \frac{25 \cdot 45^2}{29.76} = 1701 \text{ t cm}$$

$$e_m = \frac{X}{N} = \frac{1800}{27} = 66.7 \text{ cm}$$

$$c_T = \frac{d_0}{2} - d' = \frac{50}{2} - 5 = 20 \text{ cm}$$

$$e_T = e_m + c_T = 66.7 + 20 = 86.7 \text{ cm}$$

Chap. 6   Dimensioning and Ultimate Carrying Capacity

$$M_T = Ne_T = X + Nc_T = 27 \cdot 86.7 = 2341 \text{ t cm}$$

$$\Delta M = M_T - M_i = 2341 - 1701 = 640 \text{ t cm}$$

$$\Sigma A_s = \frac{M_i}{z_0 \sigma_s} + \frac{\Delta M}{c_0 \sigma_s} - \frac{N}{\sigma_s} = \frac{1701}{36 \cdot 1.400} + \frac{640}{40 \cdot 1.400} - \frac{27}{1.400}$$

$$= 33.75 + 11.43 - 19.29 = 25.88 \text{ cm}^2$$

$$\Sigma C_c = \Sigma A_s \cdot \sigma_s = 25.88 \cdot 1.400 = 36.246 \text{ t}$$

$$\Sigma z_0 = \frac{M_T}{\Sigma C_c + N} = \frac{2341}{36.246 + 27} = 37 \text{ cm}$$

Span section (Fig. 6-3):

$$x = k_x d = 0.60 \cdot 45 = 27 \text{ cm}$$

$$\frac{d - x}{x - d'} = \frac{45 - 27}{27 - 5} = 0.82$$

$$z_0 = d - t/2 = 45 - 6 = 39 \text{ cm}$$

$$= \frac{d}{0.82 \sigma_s} = \frac{45}{0.82 \cdot 1.400} = 39.2 \text{ cm}$$

$$e_m = \frac{M_m}{N} = \frac{1400}{27} = 51.9 \text{ cm}$$

$$c_B = z_B - d' = 37.3 - 5 = 32.3 \text{ cm}$$

$$e_T = e_m + c_B = 51.9 + 32.3 = 84.2 \text{ cm}$$

$$M_T = Ne_T = 27 \cdot 84.2 = 2273 \text{ t cm}$$

$$A_s = \frac{M_T}{z_0 \sigma_s} - \frac{N}{\sigma_s}$$

$$= \frac{2273}{39 \cdot 1.400} - \frac{27}{1.400} = 41.63 - 19.29 = 22.34 \text{ cm}^2$$

Diagonal tension:

$$v_1'' = \frac{V_1''}{z_0 b_0} = 11.36 \text{ kg/cm}^2$$

$$\sigma_0 = \frac{N}{A} = \frac{27000}{50 \cdot 25} = 21.6 \text{ kg/cm}$$

$$\sigma_{1,2} = -\frac{\sigma_0}{2} \pm \frac{1}{2}\sqrt{\sigma_0^2 + 4v_1''^2}$$

Combined Bending and Axial Normal Force

$$\sigma_{1,2} = -\frac{21.6}{2} \pm \frac{1}{2}\sqrt{21.6^2 + 4 \cdot 11.4^2}$$

$$= -10.8 \pm 15.7 = +4.9 \text{ kg/cm}^2 < \frac{1}{10}\sigma_{uL}$$

$$-2.5''$$

Small Eccentricity

This case applies mostly to columns subjected to combined bending and normal force. The section is subjected mainly to compressive stresses and thus the stresses in the section can be determined by the general stress equation (Eq. 6-12). However, for heavy live load ($N_L > N_D$), the σ_t may exceed the σ_c-value. In this case, the tensile stress is more than $\sigma_c/4$ of compressive stress—the large eccentricity analysis should be used (see the following example). The stress diagrams are shown in Fig. 6-5.

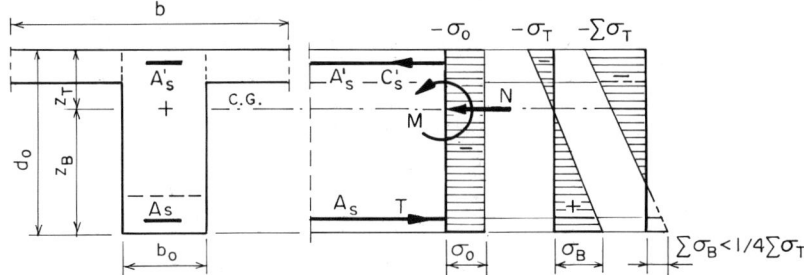

Figure 6-5

$$\sigma_T = -\frac{N}{A_c} - \frac{M}{I_0} z_T \text{—compression}$$

$$\sigma_B = -\frac{N}{A_c} + \frac{M}{I_0} z_B \text{—tension}$$

1: $\quad e_m = \dfrac{M}{N} < \dfrac{d_0}{6}$

2: $\quad e_m = \dfrac{M}{N} > \dfrac{d_0}{6}$ \hfill (6-16)

$$\frac{N}{A_c} < \frac{M}{I_0} z_B$$

$$A = d_0 b_0 \text{ cm}^2 \qquad z_T = z_B = d_0/2$$

$$I_0 = \frac{b_0 d_0^3}{12} \text{ cm}^4$$

Example

$$d_0 = 40 \text{ cm} \qquad b_0 = 30 \text{ cm} \qquad A = 1200 \text{ cm}^2$$

$$z_T = z_B = 20 \text{ cm}$$

$$I_0 = 1.6 \cdot 10^5 \text{ cm}^4$$

1: $\quad \Sigma N = 103 t \qquad M_{max} = 6.00 \text{ tm} \qquad e_m = \dfrac{600}{103} = 5.8 \text{ cm}$

$$\sigma_{T,B} = -\frac{103}{1200} \pm \frac{600}{1.6 \cdot 10^5} \, 20 = \begin{matrix} -0.010 \text{ t/cm}^2 \\ -0.160 \text{ t/cm}^2 \end{matrix}$$

2: $\quad N_D = 40 \text{ t} \qquad M_{max} = 5.00 \text{ tm} \qquad e_m = \dfrac{500}{40} = 12.5 \text{ cm}$

$$\frac{d_0}{6} = \frac{40}{6} = 6.67 \text{ cm} < e_m - \text{ tension!}$$

$$\sigma_{T,B} = -\frac{40}{1200} \pm \frac{500}{1.6 \cdot 10^5} \, 20 = \begin{matrix} +0.030 \text{ t/cm}^2 \\ -0.096 \text{ t/cm}^2 \end{matrix}$$

$$\sigma_T > \sigma_c/4 = 0.096/4 = 0.024 \text{ t/cm}^2$$

ULTIMATE CARRYING CAPACITY AND FACTOR OF SAFETY

General

The reinforced and prestressed concrete structures or structural elements are designed to carry the specified workloads. The design is based on the theory of elasticity, making use of physical characteristics and allowable stresses of the material used. The elastic theory gives sufficiently accurate results for structural analysis under working load conditions.

The ultimate carrying capacity or factor of safety traditionally is based on the ratio of critical or ultimate stresses to allowable stresses. This concept is inadequate because the ultimate strength of materials reaches into the plastic range, where stresses and deformations have no linear relationship. Therefore, determination of the ultimate carrying capacity must be based on plastic theory. Plastic theory uses ultimate stresses and strains characteristic of concrete and steel. It also assumes that plane sections remain so during bending—that is, that in the plastic range the strain is proportional to the distance from the neutral plane.

Considering the foregoing information, it is a sound practice to carry out a design in two stages: (1) working load condition based upon elastic theory, and (2) ultimate load condition based upon plastic theory.

Ultimate Carrying Capacity and Factor of Safety

Failure of a reinforced or prestressed element in bending may be due to one of the following causes:

1. Crushing of concrete
2. Rupture or excessive elongation of steel
3. Rupture of concrete in shear
4. Failure of bond or anchorage

Each of these failures is characterized by the limit strains. The element fails when limit strain of concrete or steel is reached.

Carrying capacity. The carrying capacity of an element is considered reached when the strain of extreme fiber of concrete is 0.2 percent or the steel strain due to the ultimate load (M_{uL}) reaches a maximum of 0.5 percent. The 0.5 percent steel strain is determined from the crack width: 0.1 mm for elements exposed to severe climates and up to 0.3 mm in protected conditions. In these strain limits, the element is still serviceable after the ultimate load. However, the ultimate stress limit, determined from the equilibrium requirement, must also be considered.

The analysis of failure based on the steel is relatively simple. Knowing the yield stress of steel and the lever arm (z_0) of internal forces, which varies in rather narrow limits, the carrying capacity of the element can be calculated rather accurately.

Denoting the factor of safety by F.S. = 1.8 and the working moment by M_m, the tensile force for reinforced concrete is

$$T_{uL} = \frac{1.8 \, M_m}{z_{0,uL}} \tag{6-17}$$

To determine the carrying capacity of concrete from compressive strength is rather complicated. The quality of concrete strength varies about 15 percent of the cylinder strength (σ_{uL}). Thus, the crushing strength (σ_{cr}) of concrete must be taken $\sigma_{cr} \simeq 0.85\sigma_{uL}$. The shape of the compressive stress diagram is approximately a parabola (Fig. 6-6). In accordance with tests, the ratio α of the parabolic-to-rectangular stress diagram is an average of 0.7 at crushing and the center of gravity from top is roughly $\beta \, x$ $\beta \simeq 0.4$.

The maximum compressive force based on these assumptions is

$$\Sigma C_{uL} = \alpha b x_{uL} \sigma_{cr} + n A'_s \sigma_{cr}$$

$$-\Sigma C_{uL} = T_{uL} \tag{6-18}$$

$$\sigma_{cr} = 0.85 \, \sigma_{c,uL}$$

The x-value is determined from the strain relationships of $\varepsilon_s/\varepsilon_c$. A properly designed beam often fails by crushing of the compressive zone. Therefore, the concrete strain $\varepsilon_c = 0.2$ percent controls the ultimate carrying capacity. Thus, the x-value (Fig. 6-6) is:

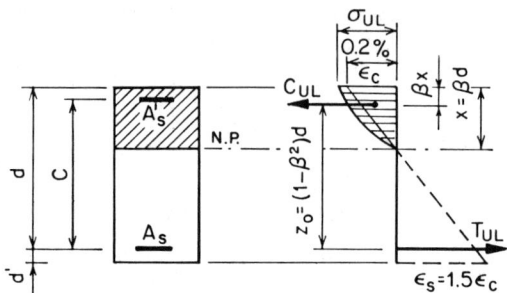

Figure 6-6

$$\varepsilon_s = \varepsilon_c \frac{d-x}{x} \qquad x = \beta d \qquad \beta = \frac{\varepsilon_c}{\varepsilon_c + \varepsilon_s} = 0.4$$

$$\varepsilon_s = \varepsilon_c \frac{1-\beta}{\beta} = \frac{1-0.4}{0.4} \varepsilon_c = 1.5\varepsilon_c \qquad (6\text{-}19)$$

$$z_0 = (1 - \beta^2)d$$

The crushing load consists of D.L + 2 L.L., $n = 15$ applies for D.L., and $n = 6$ for L.L. Taking into account the ratio D.L./L.L., the n-value for compressive reinforcing (A_s') becomes $n = 10$ under average conditions.

Example

The ultimate stress condition of the support section, discussed under large eccentricity, will be considered.

$$N = 27 \text{ t} \qquad M_T = 2341 \text{ t cm} \qquad b_0 = 25 \text{ cm}, \ d = 45 \text{ cm}$$

$$A_s = 25.88 \text{ cm}^2 \qquad A_s' = 11.43 \text{ cm}^2$$

In accordance with Eqs. 6-18 and 6-19, we obtain:

$$\sigma_{uL} = 350 \text{ kg/cm}^2 = \text{cylinder strength}$$

$$x = \beta d = 0.40 \cdot 45 = 18 \text{ cm}$$

$$z_0 = (1 - \beta^2)d = (1 - 0.4^2)45 = \sim 37.8 \text{ cm}$$

$$C_{uL} = \alpha(bx_{uL}\sigma_{cr} + nA_s'\sigma_{cr}) = 0.7\,(25 \cdot 18 \cdot 0.300 + 10 \cdot 11.43 \cdot 0.300)$$

$$= 0.7\,(135.000 + 34.290) \cong 118 \text{ t}$$

$$\sigma_{cr} = 0.85\,\sigma_{uL} = 0.85 \cdot 0.350 = 0.300 \text{ t/cm}^2$$

$$\sigma_y = \frac{T_{uL}}{A_s} \qquad T_{uL} = 25.88 \cdot 4.22 = 109.214 \text{ t} < 118 \text{ t}$$

Ultimate Carrying Capacity and Factor of Safety

Steel determines the ultimate carrying capacity, $T_{uL} < C_{uL}$

$$M_{uL} = T_{uL} \; z_0 = 109.214 \cdot 37.8 = 4128 \text{ t cm} < 1.8 \, M_T$$

$$z_0' = \frac{M_{uL}}{T_{uL}} = \frac{4128}{109.214} = 37.8 \text{ cm}$$

The shear strength is determined from the diagonal tension under ultimate load (Eq. 6-20).

$$\sigma_{1,2} = -\frac{\sigma_{cr}}{2} \pm \frac{1}{2} \sqrt{\sigma_{cr}^2 + 4v_{uL}^2} \qquad (6\text{-}20)$$

If the diagonal tension is more than $\frac{1}{10}\,\sigma_{uL}$, the total tension due to shear must be taken by stirrups. However, even if the diagonal tension is less than the tensile capacity of concrete, a minimum amount of shear reinforcing (as by code) should be provided.

The bond stress (u) is most critical for reinforced concrete. The tensile force at support is not zero, as commonly assumed, because the compressive force (C) is not horizontal but inclined (Fig. 6-7).

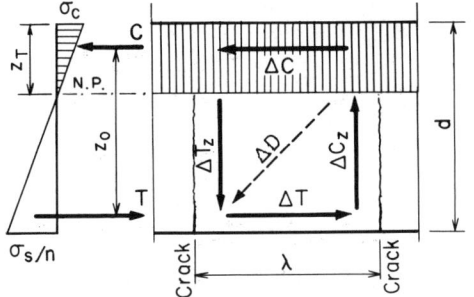
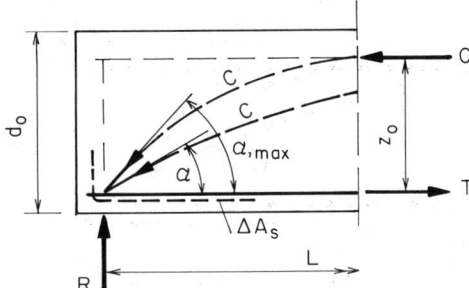

Figure 6-7(a) and (b)

$$\alpha = 20° \text{ to } 60°$$

$$T = \frac{R}{\tan \alpha} = u \cdot \Sigma 0 \qquad (6\text{-}21)$$

Therefore, the tensile reinforcing must be provided with hooks to avoid bond failure. In pretensioned elements, the strands are usually flame cut, resulting in dynamic shock, loss of bond, and cracking of end faces of beams. Conventional steel must be provided and properly designed to avoid this failure. The anchor failures for pretensioned elements will be discussed in Chapter 7.

7

Prestressed Concrete

BASIC PRINCIPLES

The basic principle of prestressing is the induction of stresses into a beam or girder before dead and live loads are applied, so that these stresses act in the opposite direction to those developed by dead and live loads. When external loads are applied, the resulting stress from loading will be superimposed by the prestress. In this manner, a more economical use of concrete and steel can be made and tensile stresses in concrete can be controlled or even entirely eliminated. Also, a major part of the dead load is carried by suspension action into supports, resulting in reduced shear.

The location of tendons and the prestressing force are determined so that stresses due to prestressing are approximately equal and opposite to those of the loading. The moments due to loading (M) and prestressing (M_{pr}) are illustrated in Fig. 7-1.

As can be seen from Fig. 7-2, in prestressed concrete the entire section is in compression, or the tensile stress is not exceeding the tensile capacity

$$\left(\frac{1}{10}\sigma_{uL}\right)$$

of concrete. Concrete behaves as homogeneous material.

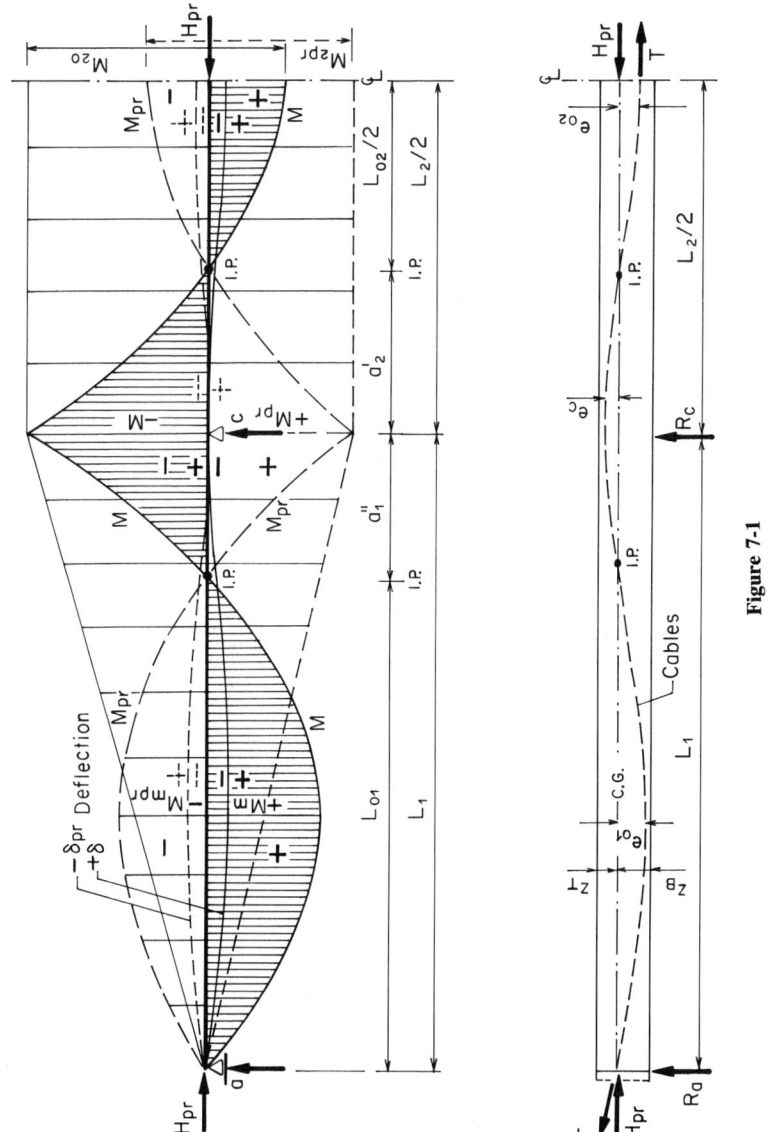

Figure 7-1

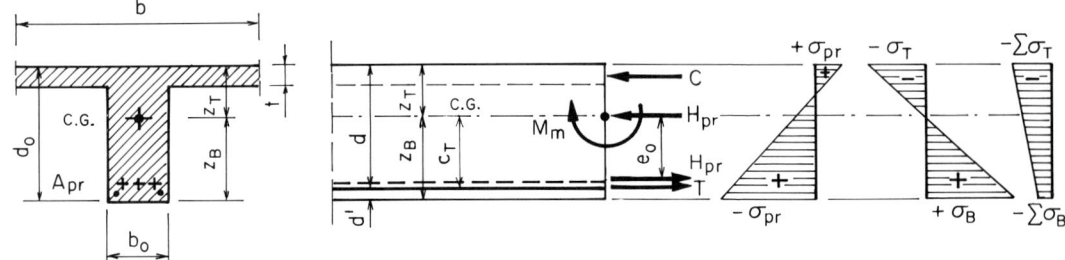

Figure 7-2

The stresses due to the external loading and prestressing are (Eq. 7-1).

$$\sigma_{T,B} = \mp \frac{M_m}{I_0} z_{T,B} \tag{7-1}$$

$$\sigma_{T,B,pr} = -\frac{H_{pr}}{A} \pm \frac{M_{pr}}{I_0} z_{T,B}$$

Considering that

$$M_{pr} = e_0 H_{pr} \qquad i^2 = \frac{I_0}{A}$$

$$\sigma_0 = -\frac{H_{pr}}{A}$$

Introducing these values into Eq. 7-1, we obtain

$$\sigma_{pr} = -\sigma_0 \left(1 \pm \frac{e_0 z_{T,B}}{i^2}\right) \tag{7-2}$$

wherein e_0 is the distance of prestressing force from the center of gravity and σ_0 is the central stress distributed uniformly over the cross-sectional area.

For simplification, we substitute the values in Eq. 7-2 with

$$1 + \frac{e_0 z_B}{i^2} = \psi_B \tag{7-3}$$

$$1 - \frac{e_0 z_T}{i^2} = \psi_T$$

where z_B is the tensile and z_T is the compressive fiber distance from the center of gravity of the section considered. Therefore, we obtain the value of fiber stresses due to the prestressing.

$$\sigma_{Bpr} = -\sigma_0 \psi_B \tag{7-4}$$

$$\sigma_{Tpr} = -\sigma_0 \psi_T$$

Thus, the prestressing force required to satisfy the Eq. 7-2 is

$$H_{pr0} = \frac{c\sigma_B}{\psi_B} A, \ (\sigma_B - \text{tensile stress}) \quad (7\text{-}5)$$

where c is the estimated loss in prestress.

APPLICATION OF PRESTRESSING

At present, there are two methods for application of prestressing: pretensioning and posttensioning.

In the pretensioning method, the prestressing strands are tensioned against exterior anchorage and in this state are embedded into fresh concrete. After the concrete has developed sufficient strength to carry the initial prestressing force (H_{pr0}), the prestressing steel is released from the external anchorage. As a result, the prestressing force is inducing compressive stresses into the concrete by bond between the prestressing steel and concrete. The pretensioning method is used for plant production, requiring heavy prestressing beds (Fig. 7-3).

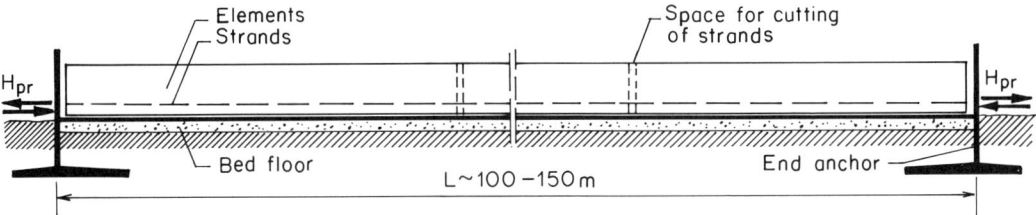

Figure 7-3

In the posttensioning method, in order to prevent bond, the prestressing tendons are placed into flexible steel or plastic tubing before they are cast into concrete, or only the steel tubing is cast into concrete and later the tendons are pulled through this tubing. When the concrete is hardened sufficiently, the tendons are tensioned by hydraulic jacks acting against the ends of the hardened concrete elements and then anchored by special devices. The tendons in steel tubing are grouted to establish bond and protection against corrosion. The posttensioning method is mostly used at the construction site for prestressed advanced structures. In bonded tendons, the prestressing cables are an integral part of concrete. In nonbonded prestressing, the tendons are covered by corrosion-preventing grease and placed into plastic tubes. In this case, the tendons act as independent structural elements.

Suspension Action

In the posttensioning method, the suspension action is dominant. There are two types of strand and cable layouts: parabolic and polygonal. The parabolic layout is used mostly in posttensioned beams and girders, as illustrated in Fig. 7-4.

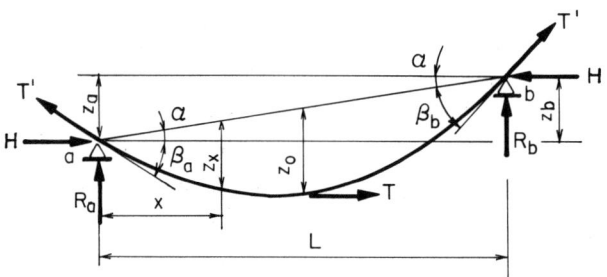

Figure 7-4

The uniform load to be carried by the cables is determined by the use of three equations of equilibrium (Eq. 1-4).

$$\Sigma Z = O: \quad R_a + R_b - wL = 0, \; \Sigma R = wL$$

$$\Sigma M = O: \quad R_a L - \frac{wL^2}{2} = 0, \; M_{\max} = \frac{1}{8} wL^2$$

$$R_a = \frac{wL}{2} = R_b$$

$$T = \frac{wL^2}{8z_0 - 4z_b} \tag{7-6}$$

$$w = \frac{T(8z_0 - 4z_b)}{L^2}$$

$$z_0 = \frac{wL^2}{8T} + \frac{z_b}{2}$$

$$\tan\alpha = \frac{z_b}{L} \qquad \tan\beta_b = \frac{wL}{2T} + \tan\alpha$$

$$\tan\beta_a = \frac{wL}{2T} - \tan\alpha$$

$$T_0' = \frac{1}{\cos\beta_a} T, \; T_b' = \frac{1}{\cos\beta_b} T$$

$$z_x = 4z_0 \, \xi\xi' \qquad \xi = \frac{x}{L} \qquad \xi' = (1 - \xi) = \frac{x'}{L}$$

Example (Fig. 7-4)

$$L = 40 \text{ m}$$

$$w = 1.0 \text{ t/m} \qquad z_b = 0.45 \text{ m}$$

$$M_0 = \frac{1}{8} 1.0 \cdot 40^2 = 200 \text{ tm}$$

$$z_0 = 1.60 \text{ m}$$

$$T = \frac{M_0}{z_0} = \frac{200}{1.60} = 125 \text{ t}$$

For relatively small angles, such as in posttensioning, $\cos \beta = 1$. Thus,

$$T' \simeq T \simeq H$$

For a symmetric layout of cables and strands, the computations are the same, except the z_b-value is zero.

The polygonal layout of cables is mostly used in posttensioned Vierendeels and trusses, and also for strengthening of existing beams and girders (Fig. 7-5).

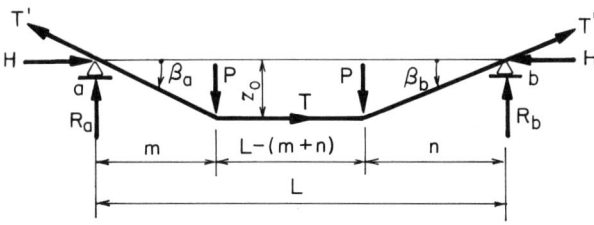

Figure 7-5

$$R_a + R_b + P_1 + P_2 = 0, \quad R_a = \frac{P_1(L - m) + P_2 n}{L}$$

$$R_0 m + T z_0 = 0 \qquad R_b = \Sigma P - R_a \qquad (7\text{-}7)$$

$$T = \frac{R_a m}{z_0}$$

STATICALLY DETERMINATE PRESTRESSED MEMBERS

Straight tendons. In straight strands, the prestressing moment (M_{pr}) and prestressing force (H_{pr}) are as illustrated in Fig. 7-6.

$$H_{pr} = \frac{c\sigma_B}{\psi_B} A \qquad -M_{pr} = H_{pr} e_0$$

As both H_{pr} and e_0 are constant, the prestressing moment is rectangular. Whatever the external moment (M_0), it is zero at the beam ends. Therefore, the superposition of the external moment (M_0) and prestressing moment (M_{pr}) cannot be balanced at the beam ends.

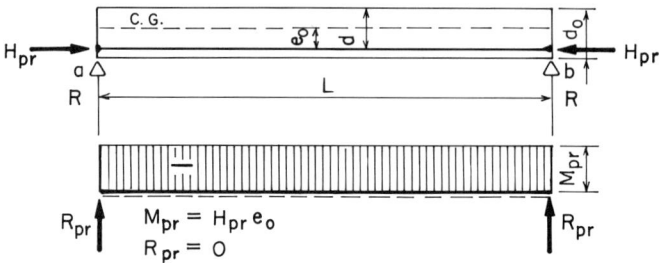

Figure 7-6

$$M_0 + M_{pr} \neq 0, (M_0 = 0)$$

These unbalanced moments (M_{pr}) cause tension at the top fiber of the beam ends and must be taken by conventional reinforcing. The H_{pr}-value is determined from the fiber stress condition (see the section "Internal Forces and Stresses").

Parabolic symmetric tendons (Fig. 7-7). In the case of parabolic tendons with end anchorage at the center of gravity of the section, the external moment due to uniform loading (M_0) and prestressing moment (M_{pr}) can be balanced along the entire length of the element so that the $M_0 - M_{pr}$ is zero and the element has only centric compressive stresses. Thus,

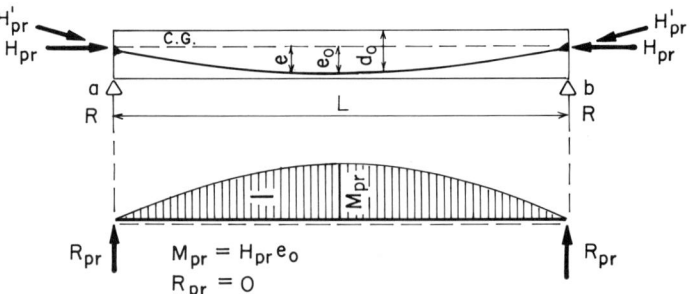

Figure 7-7

$$M_0 + M_{pr} = 0$$

$$M_{0x} = 4M_{0,max}\xi\xi'$$

$$M_{pr} = H_{pr}\, e_0 \cos\beta \qquad (7\text{-}8)$$

$$\cos\beta \sim 1.0$$

$$e_x = 4e_0\xi\xi'$$

$$\xi = \frac{x}{L} \qquad \xi' = \frac{x'}{L} = 1 - \xi$$

For both these prestressed members with straight and parabolic tendons, the reaction (R_{pr}) and shear (V_{pr}) are zero.

STATICALLY INDETERMINATE PRESTRESSED MEMBERS

For statically indeterminate systems, the statically determined elements serve as primary systems. The redundant X is the support moment due to prestressing.

Two Spans (EI = constant)

Straight tendons. The prestressing force acting in a primary system is illustrated in Fig. 7-8.

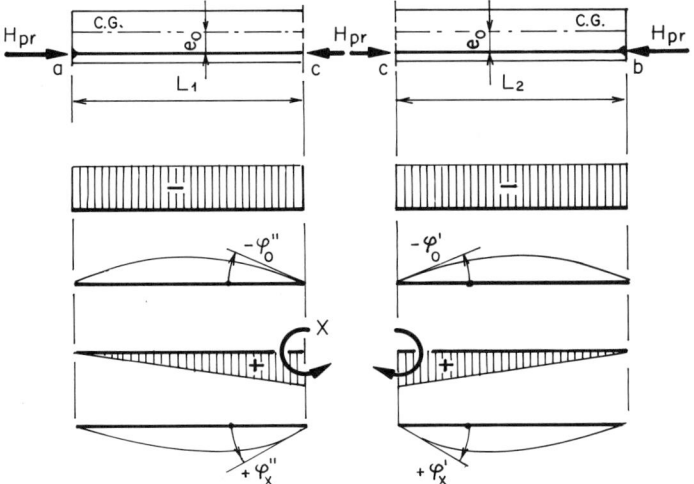

Figure 7-8

The continuity requires that the end rotations (φ) due to prestress at the support c are zero.

$$\Sigma\varphi'' + \Sigma\varphi' = 0$$

Loading factors. (see Eqs. 2-5 and 2-11):
Moment loading:

$$M_{0pr} = H_{pr}e_0 \qquad R_{0pr} = 0$$

$$EI\varphi''_{pr} = \overline{R}_1 \qquad EI\varphi'_{pr} = \overline{R}_2$$

$$\overline{L}_1'' = \frac{6}{L_1}\overline{R}_1 \qquad \overline{L}_2' = \frac{6}{L_2}\overline{R}_2$$

$$\overline{L}_1' = \overline{L}_1'' = 6\overline{R}_1 = -6\frac{H_{pr}e_aL_1}{2} = -3H_{pr}e_0L_1$$

$$\overline{L}_2' = \overline{L}_2'' = 6\overline{R}_2 = -6\frac{H_{pr}e_0L_2}{2} = -3H_{pr}e_0L_2$$

Moment loading: Redundant X

$$EI\varphi_x'' = \overline{R}_{1x} \qquad EI\varphi_x' = \overline{R}_{2x}$$

$$\overline{L}_{1x}'' = 2XL_1 \qquad \overline{L}_{2x}' = 2XL_2$$

Continuity requirement (Eq. 2-16):

$$2X(L_1 + L_2) - 3H_{pr}e_0(L_1 + L_2) = 0$$

$$X = 1.5\,H_{pr}e_0$$

Thus, prestressing moment at "c" by superposition:

$$M_{cpr} = M_{0pr} + X \qquad (7\text{-}9)$$
$$= -H_{pr}e_0 + 1.5\,H_{pr}e_0 = 0.5\,H_{pr}e_0$$

Reactions:

$$R_{apr} = \frac{M_{cpr}}{L_1} \qquad R_{bpr} = \frac{M_{cpr}}{L_2} \qquad R_{cpr} = -(R_{apr} + R_{bpr}) \qquad (7\text{-}10)$$

Moments and shear are illustrated in Fig. 7-9.

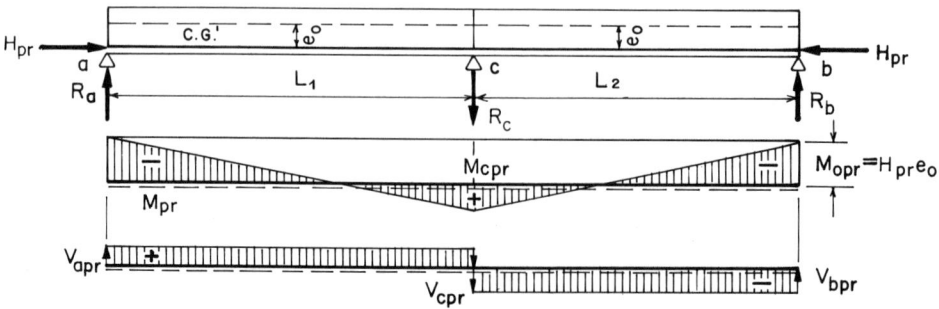

Figure 7-9

Statically Indeterminate Prestressed Members

Parabolic tendons. $e_c = 0$: (Fig. 7-10)

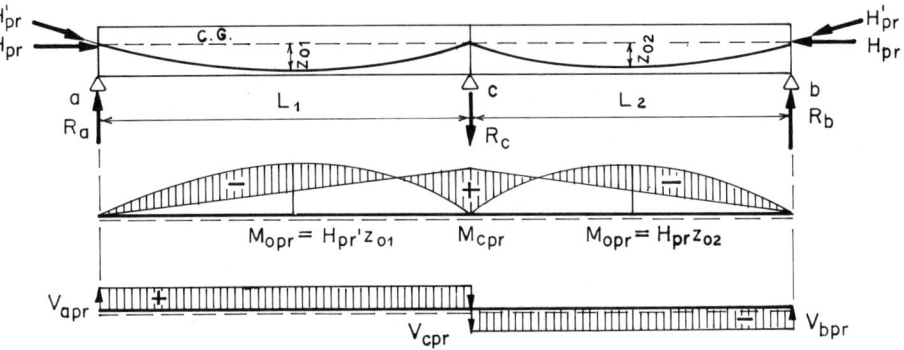

Figure 7-10

Loading factors.
Moment loadings: $M_{10pr} = H_{pr}z_{01}$ $M_{20pr} = H_{pr}z_{02}$

$$\overline{L}_1'' = -2H_{pr}z_{01}L_1$$

$$\overline{L}_2' = -2H_{pr}z_{02}L_2$$

Moment loading: Redundant X

$$\overline{L}_{1x}'' = 2XL_1$$

$$\overline{L}_{2x}' = 2XL_2$$

Continuity requirement:

$$-2H_{pr}(z_{01}L_1 + z_{02}L_2) + 2X(L_1 + L_2) = 0 \tag{7-11}$$

$$X = \frac{H_{pr}(z_{01}L_1 + z_{02}L_2)}{L_1 + L_2}$$

$$M_{c0} = 0 \tag{7-12}$$

$$z_{01} = z_{02}: \quad X = H_{pr}e_0$$

Prestressing moment at c and reactions:

$$M_{cpr} = X \tag{7-13}$$

$$R_{apr} = \frac{X}{L_1} \quad R_{bpr} = \frac{X}{L_2} \quad R_{cpr} = -(R_{apr} + R_{bpr})$$

Parabolic tendons. $e_c \neq 0$: (Fig. 7-11)

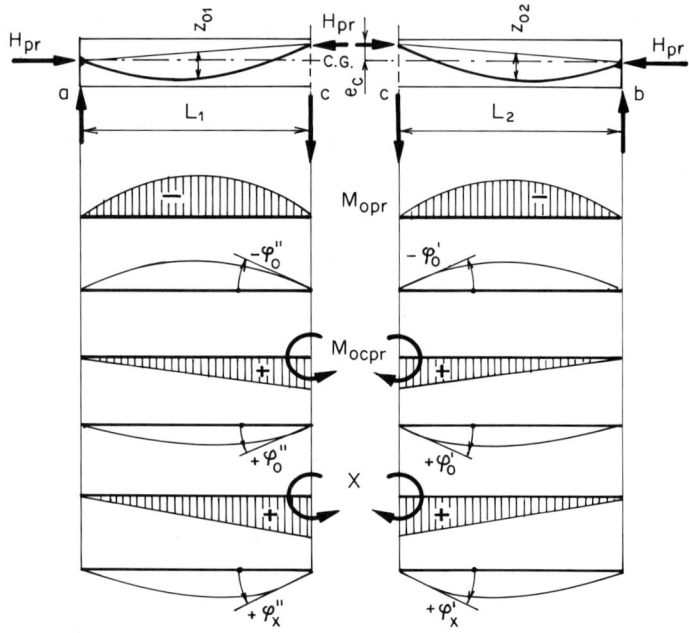

Figure 7-11

Loading factors:
$$M_{10pr} = H_{pr}z_{01} \qquad M_{20pr} = H_{pr}z_{02}$$
$$\overline{L}_1'' = -2H_{pr}z_{01}L_1 = \overline{L}_1'$$
$$\overline{L}_2' = -2H_{pr}z_{02}L_2 = \overline{L}_2''$$
$$\overline{L}_1'' = 2XL_1 \qquad \overline{L}_2' = 2XL_2$$

Continuity requirement:
$$-H_{pr}(z_{01}L_1 + z_{02}L_2) + H_{pr}e_c(L_1 + L_2) + X(L_1 + L_2) = 0 \qquad (7\text{-}14)$$
$$X = \frac{H_{pr}(z_{01}L_1 + z_{02}L_2) - e_c(L_1 + L_2)}{L_1 + L_2}$$
$$= H_{pr}\left(\frac{z_{01}L_1 + z_{02}L_2}{L_1 + L_2} - e_c\right) \qquad M_{c_0pr} = H_{pr}e_c$$
$$M_{cpr} = X + M_{c_0pr} = H_{pr}\left(\frac{z_{01}L_1 + z_{02}L_2}{L_1 + L_2}\right)$$

Statically Indeterminate Prestressed Members

$$L_1 = L_2 = L$$

$$z_{01} = z_{02} = z_0$$

$$X = H_{pr}(z_{01} - e_0)$$

By superposition, we obtain:

$$M_{cpr} = H_{pr}(z_{01} - e_c + e_c) = H_{pr}z_{01} \quad (7\text{-}15)$$

$$R_{apr} = \frac{M_{cpr}}{L_1} \qquad R_{bpr} = \frac{M_{cpr}}{L_2} \qquad R_c = -(R_{apr} + R_{bpr})$$

The prestressing moments are illustrated in Fig. 7-12.

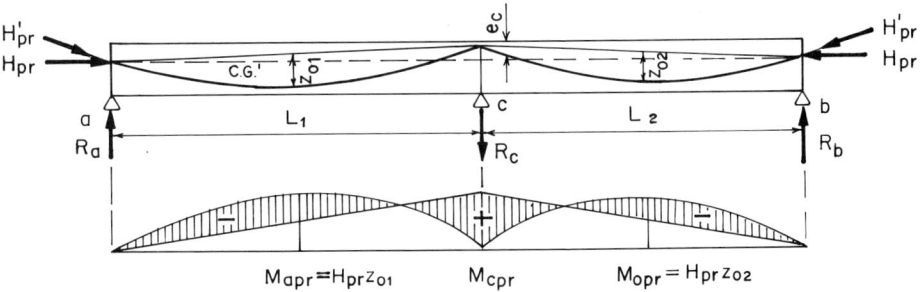

Figure 7-12

Three Spans (Fig. 7-13)

$L_1 = L_2$:

Loading factors:

$$\overline{L}_1'' = -2H_{pr}z_{01}L_1 \qquad \overline{L}_2' = -2H_{pr}z_{02}L_2 = \overline{L}_2'' \qquad \overline{L}_3' = -2H_{pr}z_{03}L_3$$

$$\overline{L}_{1c}'' = -2H_{pr}e_cL_1 \qquad \overline{L}_{2c}' = 3H_{pr}e_cL_2 = \overline{L}_2'' \qquad \overline{L}_{3c}' = 2H_{pr}e_cL_3$$

Redundants: X

$$\overline{L}_{1x}'' = 2XL_1, \overline{L}_{2x}' = 2XL_2 = \overline{L}_{2x}'' \qquad \overline{L}_{3x}' = 2XL_3$$

By continuity requirement, we obtain ($L_1 = L_3$).

$$-2H_{pr}(z_{01}L_1 + z_{02}L_2) + 2H_{pr}\left(L_1 + \frac{3}{2}L_2\right) + 2X\left(L_1 + \frac{3}{2}L_2\right) = 0$$

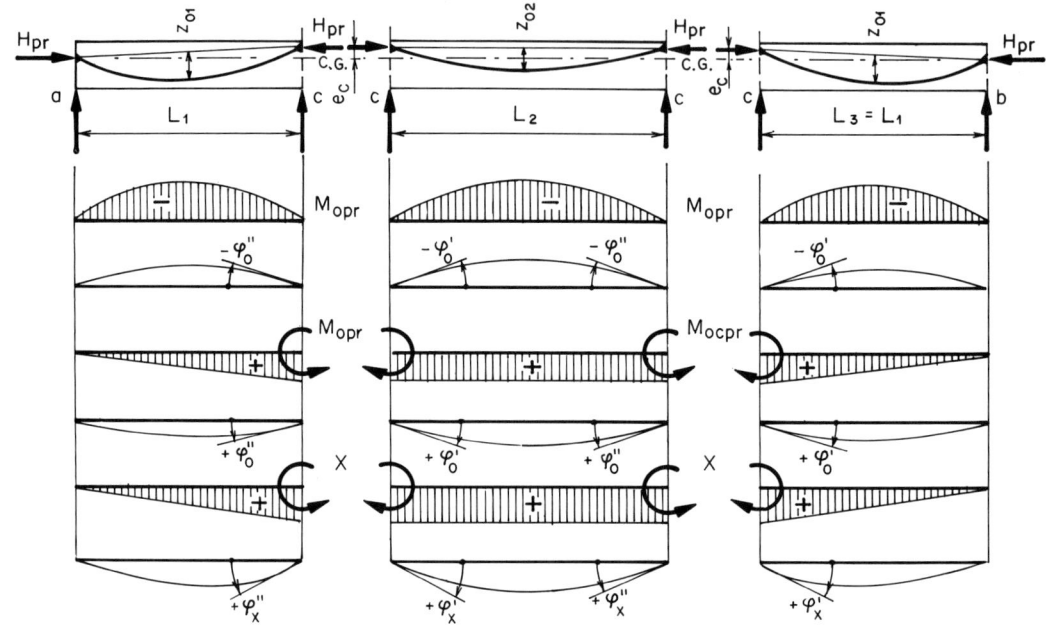

Figure 7-13

$$X = \frac{H_{pr}\left[z_{01}L_1 + z_{02}L_2 - e_c\left(L_1 + \frac{3}{2}L_2\right)\right]}{L_1 + \frac{3}{2}L_2}$$

$$M_{c_0pr} = H_{pr}e_c$$

$$M_{cpr} = X + M_{c_0pr} = H_{pr}\left(\frac{z_{01}L_1 + z_{02}L_2}{L_1 + \frac{3}{2}L_2} - e_c\right) + H_{pr}e_c$$

$$= H_{pr}\frac{z_{01}L_1 + z_{02}L_2}{L_1 + \frac{3}{2}L_2} \quad (7\text{-}16)$$

$L_1 = L_2 = L_3 = L$:

$$M_{cpr} = H_{pr}\frac{(z_{01} + z_{02})L}{\left(1 + \frac{3}{2}\right)L} = H_{pr}\frac{z_{01} + z_{02}}{1 + \frac{3}{2}} \quad (7\text{-}17)$$

$$R_{0pr} = \frac{M_{cpr}}{L_1} \qquad R_{cpr} = R_{0pr}$$

Factor of Safety

The prestressing moments are illustrated in Fig. 7-14.

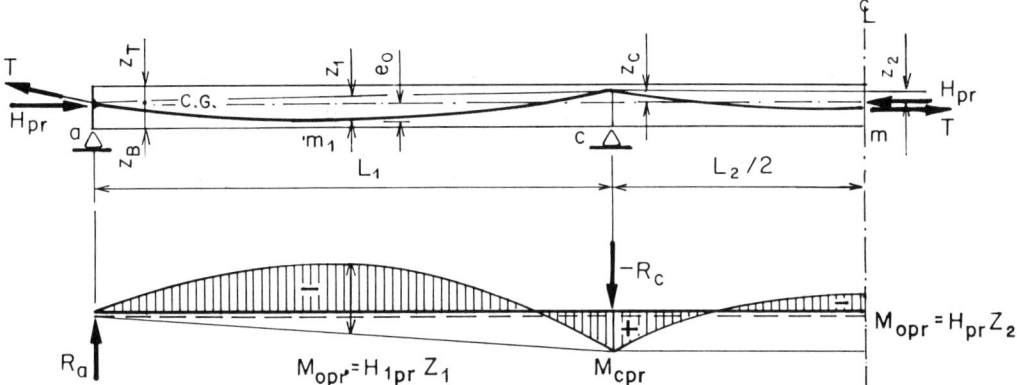

Figure 7-14

FACTOR OF SAFETY (See Chapter 6)

The ultimate carrying capacity of prestressed beams or girders is principally the same as that of bending with axial force. The prestressing force is determined from the condition that the tensile stress due to loading (σ_w) plus prestress (σ_{pr}) is

$$\sigma_w + \sigma_{pr} = 0$$

and the prestress does not exceed the elastic limit.

The starting point of the analysis of ultimate carrying capacity is $\sigma_{cw} = 0$ and the strain condition in this point is

$$\varepsilon_{pr} + \frac{n\sigma_{cpr}}{E_{pr}} = \varepsilon_{pr} + \varepsilon'_{pr}$$

The $\varepsilon_{pr} + \varepsilon'_{pr}$ has to remain below or equal to 0.8 percent.

The $\varepsilon_{pr} + \varepsilon'_{pr} -$ is the prestress used from $\varepsilon_c = 0$ to full prestress ε_{pr} and back to $\varepsilon_c = 0$. The total strain for prestressing force is

$$\Sigma\varepsilon_{uL} = \varepsilon_{pr} + \varepsilon'_{pr} \frac{1-\beta}{\beta} \varepsilon_c$$

$$\beta = 0.4 \qquad \varepsilon_c = 0.2\%$$

By using the factor of safety F.S. = 1.8, the ultimate carrying capacity is determined from the condition $\Sigma X = 0$, $\Sigma M = 0$.

$$C_{cuL} = T_{pruL} = \alpha\sigma_{uL}b\beta d \qquad \alpha = 0.70$$

$$T_{uL} = \frac{1.8\,M_w}{z_{pr}} = C_{cuL} \qquad z_{pr} = (1-\beta^2)d \qquad (7\text{-}18)$$

At the starting point $\varepsilon_c = 0$, the required tensile force is

$$T'_{pr} = \frac{M_w}{z_{pr}} = A_{pr}(\varepsilon_{pr} + \varepsilon'_{pr})E_{pr}$$

and to satisfy the safety requirements, additional tensile force is required. Thus, $E_{pr} \sim E_s$:

$$T''_{req} = \frac{0.8 M_w}{z_{pr}} = T_{uL} - T'_{pr} = \Sigma A \left(\varepsilon_{uL} - \varepsilon_{pr} - \varepsilon'_{pr} + \frac{1 - \beta}{\beta} \varepsilon_c \right) E_{pr}$$

$$\Sigma A = A_{pr} + A_s$$

The total tensile force required is

$$T_{uL} = T'_{pr} + T''_s = A_{pr}\sigma_{pr} + A_s\sigma_s$$

In case the $A_{pr}\sigma_{pr} < C_{cuL}$, conventional reinforcing is required

$$A_s = \frac{\Delta T}{\dfrac{1 - \beta}{\beta} \varepsilon_c E_{pr}} = \frac{\Delta T}{\sigma_y}$$

The stress of steel for any strain can be taken from stress-strain diagram. Very often the ε-value, to satisfy the safety requirements, may reach up to 1.1 percent.

In architectural-engineering structures, nonbonded prestress often is used. The ultimate carrying capacity required to satisfy the F.S. = 1.8 cannot be obtained without conventional steel because the prestressing force is constant. In accordance with experience, the prestressing force increases to a maximum of 5 percent (reverse friction). Thus,

$$T_{uL} = \frac{1.8 M_w}{z_{pr}} = 1.05 T_{pr} + A_s\sigma_y$$

$$\Delta T_s = T_{uL} - 1.05 T_{pr} \tag{7-19}$$

$$A_s = \frac{\Delta T_s}{\sigma_y}$$

For the T-section (span), the z_{pr} may be taken $d - \dfrac{t}{2}$, where t is the thickness of slab.

The participation of H_{pr} in ultimate carrying capacity is included in σ_{pr}-value.

For continuous beam, the overall F.S. may be taken as an average of the end-span and support sections. This is justified by redistribution of external moments in accordance with the stiffness of sections.

LOSS IN PRESTRESS

Basically, there are two main types of losses to be considered: initial loss, due to anchorage and friction; and final loss, due to shrinkage and creep. The initial losses occur during application of prestressing and the final losses after shrinkage and creep are finished ($t = n$).

Initial Losses

The anchorage loss can be directly measured for wedge-type anchorage commonly used for single-strand prestressing applications. The anchorage loss during wedge setting is about 5 mm. It can be balanced by overstressing the elongation of strands by the same amount.

The frictional losses may be estimated by the Euler-Eytelwein-Grashofschen formula

$$F = H_{pr}e^{-\mu\Sigma\alpha} \tag{7-20}$$

where μ is the friction coefficient. It may be assumed for greased strands in plastic tubing $\mu = 0.10$ and for grouted tendons in metal tubing up to $\mu = 0.30$. $\Sigma\alpha$ is the total angular change from jacking end to distance x under consideration.

The angular change of the tendons for a continuous beam are illustrated in Fig. 7-15.

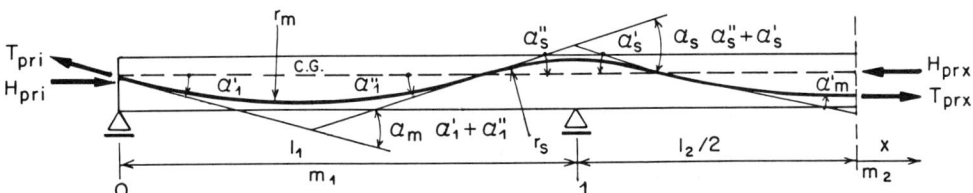

Figure 7-15

The angular change α along the tendons is normally 15° to 20° and should be kept less than 25°. The relatively small angles result in rather large radius r. Thus, the pressure $p = H_{pr}/r$ is reduced respectively and the frictional force is $F = \mu p$.

Example

$$\alpha'_m = 10° \quad \alpha''_m = 15° \quad \alpha_m = \alpha'_m + \alpha''_m = 25°$$

$$\alpha''_1 = 18° \quad \alpha'_2 = 10° \quad \alpha_s = \alpha''_1 + \alpha'_2 = 28°$$

$$\alpha'_{m2} = \alpha''_{m2} = 10°$$

Thus, the frictional loss at m_2 is: $\Sigma\alpha = 63° = 1.10 = \dfrac{\Sigma\alpha \cdot 2\pi}{360°}$

$$F_{m2} = H_{pri}e^{-\mu\Sigma\alpha} \qquad e = 2.781 \qquad \mu = 0.1$$

$$= 4 \cdot 15 \cdot 2.781^{-0.1 \cdot 1.10} = 60 \cdot 0.0896 \cong 5.4 \text{ t}$$

In accordance with experience, the $e^{-\mu\Sigma\alpha}$-value can be lightly reduced by slower or represstressing application. By doing so, the loss due to the creep of steel (~2 percent) can also be disregarded.

Loss in Prestress Due to Shrinkage and Creep*

In architectural engineering for prestressed structures, nonbonded tendons are commonly used. This means that the tendons are not an integral part of the section but are acting as an external force (H_{pr}). Therefore, the loss in prestress (ΔH_{prn}) by shrinkage and creep causes shortening of the beam. This can be estimated by Eq. 7-21.

$$\Delta H_{prn} = H_{prc}\left(1 + \frac{\varepsilon_{sr}}{\varphi}S_c\right)(1 - e^{-\alpha\varphi}) \qquad (7\text{-}21)$$

A beam subjected to prestressing force (H_{pr}) consists of concrete and reinforcing steel. Since steel is not subjected to plastic deformation (creep and shrinkage), its part of the carrying action can be determined from the strain stiffness of steel S_s and concrete S_c.

$$S_s = E_s A_s$$

$$S_c = E_0 A_c = E_0 \frac{H_{pr0}}{\sigma_{0pr}} \qquad A_c = \frac{H_{pr0}}{\sigma_{0pr}} \qquad \sigma_{0pr} = \frac{H_{pr0}}{A_c} \qquad (7\text{-}22)$$

$$\alpha = \frac{E_s A_s}{E_s A_s + E_0 A_c}$$

$$S = \frac{E_s}{E_0}\frac{A_s}{A_c} = np \qquad n = 6 \qquad p = A_s/A_c$$

Substituting $E_s A_s = S E_0 A_c$ for $E_s A_s$ in Eq. 7-22, we obtain

$$\alpha = \frac{S}{S+1} = \frac{np}{np+1} \qquad (7\text{-}23)$$

The prestressing force (H_{pr}) often is determined for full loading ($\Sigma w = w_D + w_L$), but the loss of prestress is caused by sustained load ($w_s = w_D + 0.25\,w_L$). Thus, the concrete part of the carrying action to be considered for determining the loss in prestress is

---

*August E. Komendant, *Prestressed Concrete Structures* (New York: McGraw-Hill Book Co., 1952), and *Contemporary Concrete Structures* (New York: McGraw-Hill Book Co., 1972).

Loss in Prestress

$$H_{prc} = H_{pr0}(1 - \alpha), \qquad \omega = \frac{w_s}{\Sigma w} \qquad (7\text{-}24)$$

The influence of creep in plane concrete is expressed by the ratio of plastic (ε_{pl}) to elastic (ε_{el}) strain

$$\varphi = \frac{\varepsilon_{pl}}{\varepsilon_{el}}$$

The reinforcing steel reduces the φ-value by the factor $n_n p$. Thus,

$$\varphi' = \varphi(1 - n_n p) \qquad (7\text{-}25)$$

Assuming the shrinkage proportional to creep and expressing it as a ratio of shrinkage to elastic strain, and considering that the plastic deformations obey Hooke's law, we obtain the shrinkage influence

$$\frac{\varepsilon_{sr}}{\varphi} S_c = \frac{\varepsilon_{sr}}{\varphi} \cdot \frac{E_0 H_{pr0}}{\sigma_{0pr}}$$

Introducing the values of H_{pr0}, ω, φ', and shrinkage influence into Eq. 7-21, we obtain the loss in prestressing force (ΔH_{pr}) after the plastic deformations are almost finished. Thus, $t = n$:

$$\Delta H_{pr} = H_{pr0}\left(\omega + \frac{\varepsilon_{sr}}{\varphi}\frac{E_0}{\sigma_{0pr}}\right)(1 - e^{-\alpha\varphi'}) \qquad (7\text{-}26)$$

In accordance with the author's experience and tests carried out at the construction site on posttensioned beams and girders in place, the loss in prestressing force (ΔH_{pr}), as computed by Eq. 7-26, is rather accurate.

In prestressed frames, the resistance of columns to prestressing force can be considerable and it should not be overlooked when rigid joints between beams and columns are used. However, by the use of elastically controlled joints (Fig. 3-10), the column resistance can be reduced to a fraction of that of rigid joints.

Example

A three-span continuous beam, having equal spans of $L = 15.00$ m and subjected to uniform load of 1.50 t/m (Fig. 7-16), is to be designed as a posttensioned continuous beam.

Statical computations (see Table 1 in Appendix):

$$L_1 = L_2 = L_3 = L = 15.00 \text{ m}$$

$$w = 1.50 \text{ t/m}$$

$$X_1 = X_2 = [-(0.0625 + 0.0500) + 0.0156]wL^2$$

$$= -0.0969 \cdot 1.50 \cdot 15.00^2 = -32.70 \text{ tm}$$

$$M_0 = \frac{1}{8} wL^2 = \frac{1}{8} 1.50 \cdot 15.00^2 = 42.19 \text{ tm}$$

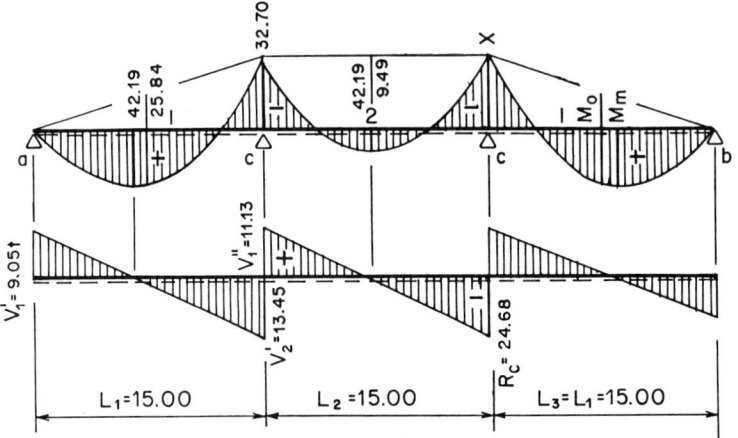

Figure 7-16

$$M_{1m} = M_0 - \frac{X}{2} = 42.19 - \frac{32.70}{2} = 25.84 \text{ tm}$$

$$M_{2m} = M_0 - X = 42.19 - 32.70 = 9.49 \text{ tm}$$

$$V'_1 = R_0 = 0.4 \cdot 1.50 \cdot 15.00 = 9.00 \text{ t} = R_3$$
$$V''_1 = 0.6 \cdot 1.50 \cdot 15.00 = 13.50 \text{ t} \quad | \quad R_1 = 24.75 \text{ t} = R_2$$
$$V'_2 = 0.5 \cdot 1.50 \cdot 15.00 = 11.25 \text{ t}$$

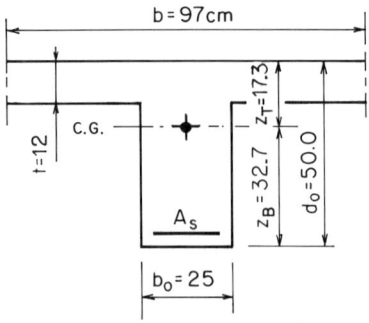

Figure 7-17

$$d_0 = \frac{1}{30} L = 5.0 \text{ dm}, \ b_0 = 2.5 \text{ dm}, \ t = 12 \text{ cm} = 1.2 \text{ dm}$$

$$b = 6t + b_0 = 6 \cdot 12 + 25 = 97 \text{ cm} = 9.7 \text{ dm}$$

$$b' = b - b_0 = 72 \text{ cm} = 7.2 \text{ dm}$$

Loss in Prestress

$$A = 2.5 \cdot 5.0 = 12.50 \text{ dm}^2 \qquad Q_T = 12.50 \cdot 2.50 = 31.25 \text{ dm}^3$$

$$1.2 \cdot 7.2 = 8.64 \text{ dm}^2 \qquad 8.64 \cdot 0.60 = 5.19 \text{ dm}^3$$

$$\Sigma A = 21.14 \text{ dm}^2 \qquad \Sigma Q_T = 36.44 \text{ dm}^3$$

$$z_T = \frac{Q_T}{A} = \frac{36.44}{21.14} = 1.73 \text{ dm} = 17.3 \text{ cm}$$

$$z_B = d_0 - z_T = 50.0 - 17.3 = 32.7 \text{ cm} = 3.27 \text{ dm}$$

$$I_T = \frac{1}{3} b_0 d_0^3 + \frac{1}{3} b' t^3 = \frac{1}{3} \cdot 2.5 \cdot 5.0^3 + \frac{1}{3} \cdot 7.2 \cdot 1.2^3$$

$$= 104.20 + 4.14 = 108.34 \text{ dm}^4$$

$$I_0 = I_T - z_T^2 A = 108.34 - 1.73^2 \cdot 21.14 = 45.07 \text{ dm}^4 = 45.07 \cdot 10^4 \text{ cm}^4$$

$$i^2 = \frac{I_0}{A} = \frac{45.07 \cdot 10^4}{21.14 \cdot 10^2} = 213 \text{ cm}^2$$

Span 1:

$$\sigma_{1T} = -\frac{M_{1m}}{I_0} z_T = -\frac{25.84}{45.07 \cdot 10^4} 17.3 = -0.098 \text{ t/cm}^2 = -98 \text{ kg/cm}^2$$

$$\sigma_{1B} = \frac{M_{1m}}{I_0} z_B = \frac{25.84}{45.07 \cdot 10^4} 32.7 = +0.189 \text{ t/cm}^2 = +189 \text{ kg/cm}^2$$

Span 2:

$$\sigma_{2T} = -\frac{M_{2m}}{I_0} z_T = -\frac{949}{45.07 \cdot 10^4} 17.3 \cong -0.036 \text{ t/cm}^2 = -36 \text{ kg/cm}^2$$

$$\sigma_{2B} = -\frac{M_{2m}}{I_0} z_B = \frac{949}{45.07 \cdot 10^4} 32.7 \cong 0.069 \text{ t/cm}^2 = 69 \text{ kg/cm}^2$$

Support section c

$$\sigma_T = \frac{X}{I_0} z_T = \frac{32.70}{45.07 \cdot 10^4} 17.3 = 0.125 \text{ t/cm}^2 = 125 \text{ kg/cm}^2$$

$$\sigma_B = -\frac{X}{I_0} z_B = -\frac{32.70}{45.07 \cdot 10^4} 32.7 = -0.237 \text{ t/cm}^2 = 237 \text{ kg/cm}^2$$

Prestressing force: Support section c: (critical)

$$e_0 = \frac{z_1 + z_2}{1 + 3/2} = \frac{33.8 + 12.0}{2.5} = 18.3 \text{ cm (Eq. 7-17)}$$

$$\psi_T = \left(1 + \frac{e_0 z_T}{i^2}\right) = 1 + \frac{18.3 \cdot 17.3}{213} = 2.486$$

$$\psi_B = \left(1 - \frac{e_0 z_B}{i^2}\right) = 1 - \frac{18.3 \cdot 32.7}{213} = -1.810$$

Expected loss in prestress is estimated as 15 percent ($c = 1.15$). Thus,

$$H_{pr0} = \frac{c\sigma_T}{\psi_T} A = \frac{1.15 \cdot 0.125}{2.486} 2114 = {\sim}123 \text{ t}$$

Use: $\quad H_{pr0} = 8 \cdot 15 \text{ t} = 120 \text{ t}$

Layout of strands is indicated in Fig. 7-18.

From span 1, six strands—$H'_{pr1} = 6 \cdot 15 = 90$ t

From span 2, two strands—$H'_{pr2} = 2 \cdot 15 = 30$ t

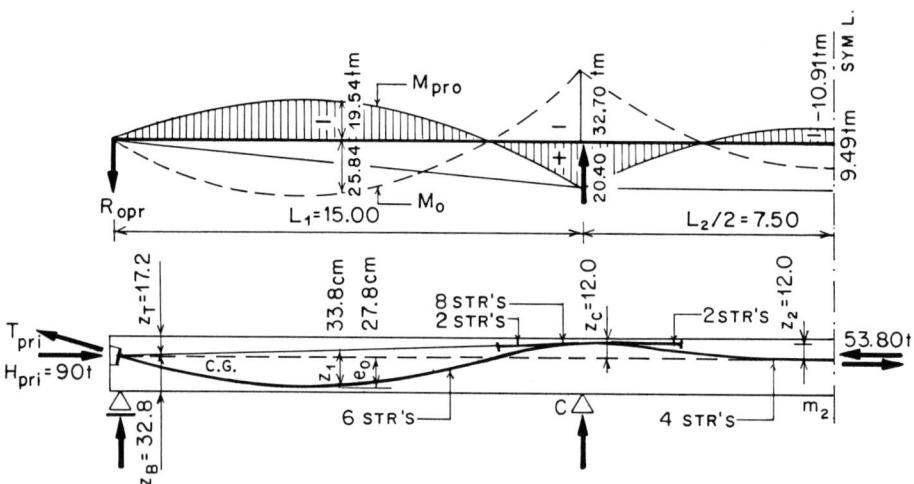

Figure 7-18

Frictional coefficient $\mu = 0.10$:

$$\alpha_{m1} = 10° + 15° = 0.17 + 0.26 = 0.43$$

$$\alpha_c = 18° + 10° = 0.31 + 0.17 = 0.48$$

$$\alpha_{m2} = 6° + 6° = 0.10 + 0.10 = 0.20$$

In accordance with Eq. 7-20, frictional losses:

$$H_{pr1} - F_{m1} = H_{pr1}(1 - e^{-\mu\alpha}) = 90(1 - 2.781^{-0.10 \cdot 0.17}) \cong 90 - 2.0 = 88 \text{ t}$$

$$H_{prc} - F_c = H_{prc}(1 - e^{-\mu\Sigma\alpha}) = 120(1 - 2.781^{-0.10 \cdot 0.74}) \cong 120 - 8.6 = 111.4 \text{ t}$$

$$H_{pr2} - F_{m2} = H_{pr2}(1 - e^{-\mu\Sigma\alpha}) = 60(1 - 2.781^{-0.10 \cdot 1.01}) \cong 60 - 6.2 = 53.8 \text{ t}$$

Loss in Prestress

Stresses:
Section c:
$t = 0$

$$M_{cpr0} = H_{cpr}\left(\frac{z_1 + z_2}{2.5}\right) = (120 - 8.6)\frac{33.8 + 12.0}{2.5} = 2040 \text{ t cm}$$

$$\sigma_{0pr} = \frac{H_{pr0}}{A_c} = -\frac{111.4}{2114} = -0.053 \text{ t/cm}^2$$

$$\sigma_{Tpr} = \sigma_0 \psi_T = -0.053 \cdot 2.486 = -0.132 \text{ t/cm}^2$$

$$\sigma_{Bpr} = \sigma_0 \psi_B = 0.053 \cdot 1.810 = 0.096 \text{ t/cm}^2$$

$$\Sigma \sigma_T = \sigma_T + \sigma_{Tpr} = 0.125 - 0.132 = -0.007 \text{ t/cm}^2$$

$$\Sigma \sigma_B = \sigma_B + \sigma_{Bpr} = -0.237 + 0.096 = -0.141 \text{ t/cm}^2$$

Section m1:

$$M_{pr0} = -19.54 \text{ tm}, \quad H_{pr0} = 88.00 \text{ t}$$

$$\sigma_{Tpr} = -\frac{H_{pr0}}{A} + \frac{M_{pr0}}{I_0} z_T = -\frac{88.00}{2114} + \frac{1954}{45.07 \cdot 10^4} 17.3$$
$$= -0.042 + 0.075 = 0.033 \text{ t/cm}^2$$

$$\sigma_{Bpr} = -\frac{88.00}{2114} - \frac{1954}{45.07 \cdot 10^4} 32.7 = -0.042 - 0.142 = -0.184 \text{ t/cm}^2$$

$$\Sigma \sigma_T = \sigma_T + \sigma_{Tpr} = -0.098 + 0.033 = -0.065 \text{ t/cm}^2$$

$$\Sigma \sigma_B = \sigma_B + \sigma_{Bpr} = 0.189 - 0.184 = 0.005 \text{ t/cm}^2$$

Section m2:

$$e_0 = 0.0, \quad H_{pr0} = -53.80 \text{ t} \quad M_{pr0} = -10.91 \text{ tm}$$

$$\sigma_{0pr} = \frac{H_{pr0}}{A} = -\frac{53.8}{2114} = -0.025 \text{ t/cm}^2$$

$$\Sigma \sigma_T = \sigma_T + \sigma_{pr} = -0.036 - 0.025 + \frac{1091}{45.07 \cdot 10^4} 17.3 = -0.019 \text{ t/cm}^2$$

$$\Sigma \sigma_B = \sigma_B - \sigma_{pr} = 0.069 - 0.025 - \frac{1091}{45.07 \cdot 10^4} 32.7 = -0.035 \text{ t/cm}^2$$

Reactions:

$$R_1' \cong 0.4 \, wL_1 = 0.4 \cdot 1.50 \cdot 15.00 = 9.000 \text{ t}$$

$$R_1'' = 0.6 \, wL_1 = 0.6 \cdot 1.50 \cdot 15.00 = 13.500 \text{ t}$$

$$R_2' = 0.5\, wL_2 = 0.5 \cdot 1.50 \cdot 15.00 = 11.250 \text{ t} = R_2''$$

$$R_{1pr}' = \frac{M_{cpr}}{L_1} = -\frac{20.39}{15.00} = -1.360 \text{ t} = +R_1''$$

$$R_{2pr}' = R_{2pr}'' = 0$$

$$\Sigma R_1' = R_1' - R_{1pr} = 9.000 - 1.360 = 7.640 \text{ t}$$

$$\Sigma R_1'' = R_1'' + R_{1pr} = 13.50 + 1.360 = 14.860 \text{ t}$$

$$\Sigma\Sigma R_1 = R_1'' + R_2' = 24.75 \text{ t}$$

Shear:

$$V_1'' = \frac{\Sigma R_1''}{b_0 z_0} = \frac{14.86}{25 \cdot 40} = 0.015 \text{ t/cm}^2$$

$$\sigma_{0pr} = 0.053 \text{ t/cm}^2, \text{ (Section c)}$$

Diagonal tension:

$$\sigma_{1,2} = -\frac{\sigma_{0pr}}{2} \pm \frac{1}{2}\sqrt{\sigma_{0pr}^2 + 4V_1''^2}$$

$$= -\frac{0.053}{2} \pm \frac{1}{2}\sqrt{0.053^2 + 4 \cdot 0.015^2} = \begin{matrix} +0.004 \text{ t/cm}^2 \\ -0.057 \text{ t/cm}^2 \end{matrix}$$

Anchorage stress:* 6-12.5mm strands = 90t

$$\sigma_{A,uL} = 0.85\sigma_{uL}\sqrt{\frac{A_L}{A_A}} = 0.85 \cdot 350 \sqrt{\frac{17 \cdot 44}{12 \cdot 39}} = 376 \text{ kg/cm}^2$$

$$\sigma_{A,AL} = 0.60\sigma_{A,uL} = 0.60 \cdot 376 = 226 \text{ kg/cm}^2$$

$$A_A = \frac{H_{pr,0}}{\sigma_{A,AL}} = \frac{90 \cdot 1000}{226} = 398 \text{ cm}^2 < 12 \cdot 39 = 468 \text{ cm}^2$$

$$\sigma_{Ay} = \nu\sigma_A = 0.167\, \frac{90000}{468} = 32 \text{ kg/cm}^2 - \text{tension}$$

Factor of safety and conventional steel:
Span 1:

$$M_{1uL} = FS \cdot M_{m1} = 1.8 \cdot 25.84 = 46.51 \text{ tm}$$

*August E. Komendant, *Prestressed Concrete Structures* (New York: McGraw-Hill Book Company, Inc., 1952).
K. H. Middendorf, "Practical Aspects of End Zone Bearing of Post-Tensioning Tendons," *Journal of the Prestressed Concrete Institute,* Volume 8, Nr4, (August 1963).

Loss in Prestress

$$T_{max} = \frac{M_{1uL}}{z_0} = \frac{46.51}{37.8} = 123 \text{ t}, \; z_0 = (1 - 0.4^2)d$$

$$\Delta T = 123 - 1.05 \cdot 88 = 31 \text{ t}$$

$$A_s = \frac{\Delta T}{\sigma_y} = \frac{31000}{4000} = 7.75 \text{ cm}^2$$

Support section c:

$$M_{c,uL} = 1.8 \cdot 32.70 = 58.86 \text{ tm}$$

$$T_{uL} = \frac{M_{cuL}}{z_0} = \frac{5886}{37.8} \cong 156 \text{ t}$$

$$\Delta T_c = 156 - 1.05 \cdot 111.4 = 39 \text{ t}$$

$$A_s = \frac{\Delta T_c}{\sigma_y} \sim \frac{39000}{4000} \cong 9.75 \text{ cm}^2$$

Loss Due to Shrinkage and Creep

$$E_s = 21.0 \cdot 10^5 \text{ kg/cm}^2$$

$$E_0 = 3.5 \cdot 10^5 \text{ kg/cm}^2$$

$$n = \frac{E_s}{E_0} = 6 \qquad n_n = 15$$

$$\omega = \frac{w_s}{\Sigma w} \cong 0.80 \qquad \sigma_{0pr} = \frac{H_{prc}}{A_c}$$

$$p_c = \frac{A_s}{A_c} = \frac{2 \cdot 10.2}{2114} = 0.0096 \qquad p_{m1} = 0.078$$

$$\varepsilon_{sr} = 20.0 \cdot 10^{-5}$$

$\varphi = 2$, φ-value depends on the quality of concrete, and relative humidity may vary from 1.6 to 3.0.

Support section c:

$$\alpha = \frac{np_c}{np_{c+1}} = \frac{6 \cdot 0.0096}{6 \cdot 0.0096 + 1} = 0.055$$

$$H_{prc} = H_{pro}(1 - \alpha) = 111.4(1 - 0.055) = 105.3 \text{ t}$$

$$\varphi' = \varphi(1 - n_n p_c) = 2(1 - 15 \cdot 0.0096) = 1.712$$

$$\Delta H_{pr} = H_{prc}\left(\omega + \frac{\varepsilon_{sr}}{\varphi} \cdot \frac{E_0}{\sigma_{0pr}}\right)(1 - e^{-\alpha\varphi'}), \quad \sigma_{0pr} = 50 \text{ kg/cm}^2$$

$$= 105.3\left(0.80 + \frac{20 \cdot 10^{-5}}{2} \cdot \frac{3.5 \cdot 10^5}{50}\right)(1 - 2.781^{-0.055 \cdot 1.712})$$

$$= 105.3 \cdot 1.50 \cdot 0.09 = 14.22 \text{ t}$$

$$\Delta\sigma_{0pr} = \frac{\Delta H_{pr}}{A_c} = \frac{14.22}{2114} = 0.0067 \text{ t/cm}^2$$

$$\Delta\sigma_{Tpr} = \Delta\sigma_{0pr}\psi_T = -0.0067 \cdot 2.486 \cong -0.017 \text{ t/cm}^2$$

$$\Delta\sigma_{Bpr} = \Delta\sigma_{0pr}\psi_B = -0.0067 \cdot 1.810 = -0.012 \text{ t/cm}^2$$

$$\Sigma\sigma_{Tn} = \Sigma\sigma_{T0} - \Delta\sigma_{Tpr}$$
$$= -0.007 + 0.017 = +0.010 \text{ t/cm}^2 = 10 \text{ kg/cm}^2$$
$$\Sigma\sigma_{Bn} = \Sigma\sigma_{B0} - \Delta\sigma_{Bpr}$$
$$= -0.141 - 0.012 = -0.153 \text{ t/cm}^2 = -153 \text{ kg/cm}^2$$

Span section m1:

$$\alpha = \frac{np_m}{np_m + 1} = \frac{6 \cdot 0.0078}{6 \cdot 0.0078 + 1} = 0.045$$

$$H_{prc} = H_{pr0}(1 - \alpha) = 88(1 - 0.045) = 84 \text{ t}$$

$$\sigma_{0pr} = \frac{H_{prc}}{A_c} = \frac{84 \cdot 1000}{2114} = 40 \text{ kg/cm}^2$$

$$\varphi' = \varphi(1 - n_n p_m) = 2(1 - 15 \cdot 0.0078) = 1.770$$

$$\Delta H_{prm} = H_{prc}\left(\omega + \frac{\varepsilon_{sr} \cdot 10^{-5}}{\varphi} \cdot \frac{E_0}{\sigma_{0pr}}\right)(1 - e^{-\alpha\varphi'})$$

$$= 84\left(0.80 + \frac{20.0 \cdot 10^{-5}}{2} \cdot \frac{3.5 \cdot 10^5}{40}\right)(1 - 2.781^{-0.045 \cdot 1.770})$$

$$= 84 \cdot 1.675 \cdot 0.080 = 11.26 \text{ t}$$

$$\Delta\sigma_{0pr} = \frac{\Delta H_{pr}}{A_c} = \frac{11.26}{2114} = 0.005 \text{ t/cm}^2 = 5 \text{ kg/cm}^2$$

Deflection

$$\Delta M_{mpr} = \Delta H_{prm} z_1 - \Delta H_{prc} \frac{z_1 + z_c}{2 \cdot 2.5}$$

$$= 11.24 \cdot 33.8 - 14.22 \frac{33.8 + 12.0}{5} = 380 - 130 = 250 \text{ t cm}$$

$$\Delta \sigma_{Tpr} = \frac{\Delta H_{pr}}{A_c} \pm \frac{\Delta M_{prm}}{I_0} z_T = -\frac{11.24}{2114} + \frac{250}{45.07 \cdot 10^4} 17.3 = \sim 0.005 \text{ t/cm}^2$$

$$\Delta \sigma_{Bpr} = -0.005 - \frac{250}{45.07 \cdot 10^4} \cdot 32.7 = -0.023 \text{ t/cm}^2$$

$$\Sigma \sigma_{Tn} = \Sigma \sigma_{T0} + \Delta \sigma_{Tpr} = -(0.065 + 0.005) = -0.070 \text{ t/cm}^2$$
$$= -70 \text{ kg/cm}^2$$

$$\Sigma \sigma_{Bn} = \Sigma \sigma_{B0} - \Delta \sigma_{Bpr}$$
$$= +0.005 + 0.023 = 0.028 \text{ t/cm}^2 < 0.035 \text{ t/cm}^2$$

Stresses for $t = 0$ and $t = n$ are illustrated in Fig. 7-19.

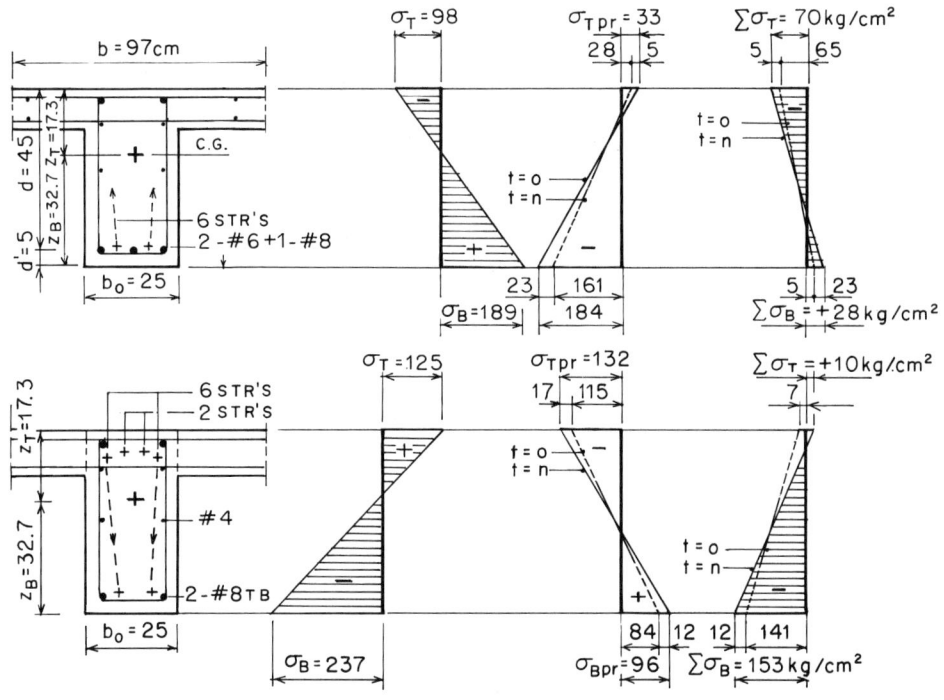

Figure 7-19

DEFLECTION

The deflections of span L_1 (critical) can be obtained by superposition from deflection due to M_{m1} and M_{pr1} (Fig. 7-20).

$$\delta_{m1} = \sum_D^L \delta_m - \delta_{pr1}$$

Dead and live load:

$$\Sigma M_m = 42.19 \text{ tm} \qquad X_1 = -32.70 \text{ tm} \qquad x = \frac{L_1}{2}$$

$$E_0 I_0 \delta_{m0} = \frac{5 \cdot 8}{384} \Sigma M_{m1} L_1^2 = 0.104 \cdot 42.19 \cdot 15^2 = +988 \text{ tm}^3$$

$$E_0 I_0 \delta_X = -\frac{X_1}{6}\left(L_1 - \frac{x}{L_1}\right)x = -\frac{32.70}{6}(15 - 0.5)\frac{L_1}{2} = -592 \text{ tm}^3$$

$$E_0 I_0 \delta_{m1} = 988 - 592 = +396 \text{ tm}^3$$

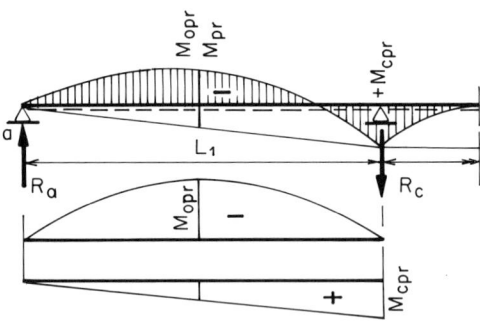

Figure 7-20

Prestressing:

$$\Sigma M_{pr} = z_1 H_{pr} = 0.338 \cdot 88 = -29.74 \text{ tm}$$

$$X_{pr} = 20.40 \text{ tm}$$

$$E_0 I_0 \delta_m = -0.104 \cdot 29.74 \cdot 15^2 = -696 \text{ tm}^3$$

$$E_0 I_0 \delta_X = \frac{20.40}{6} \cdot 14.5 \cdot 15/2 = 370 \text{ tm}^3$$

Deflection

$t = 0$:

$$E_0 I_0 \delta_{pr} = -696 + 370 = -326 \text{ tm}^3$$

$$E_0 I_0 = 350 \cdot 45.07 \cdot 10^4 = 157.74 \cdot 10^6 \text{ tcm}^2$$

$$\delta_m = \frac{326 \cdot 10^6}{157.74 \cdot 10^6} = 2.5 \text{ cm}$$

$$\delta_{pr} = -\frac{202 \cdot 10^6}{157.74 \cdot 10^6} = -2.1 \text{ cm}$$

$$\Sigma\Sigma\delta_{m0} = 2.5 - 2.1 = 0.4 \text{ cm-downwards}$$

$t = n$:

$$H_{pn}/H_{pr0} = \frac{111.4 - 14.22}{111.4} = \frac{88.0 - 11.26}{88} = 0.87$$

$$E_0/E_n = \frac{3.5 \cdot 10^5}{2.0 \cdot 10^5} = 1.75 \qquad E_n \cong \frac{E_0}{\varphi'} \cong 2.0 \cdot 10^5 \qquad \omega = 0.80$$

$$\Sigma\Sigma\delta_{m,n} = 2.5 \cdot 1.75 \cdot 0.80 - 2.1 \cdot 0.87 = 1.7 \text{ cm downwards}$$

For simplicity and economy, the Vierendeels are commonly manufactured in horizontal position. Therefore, Gerber-type continuous Vierendeels are used. One of such posttensioned Vierendeels with end diagonals is illustrated in Fig. 7-21.

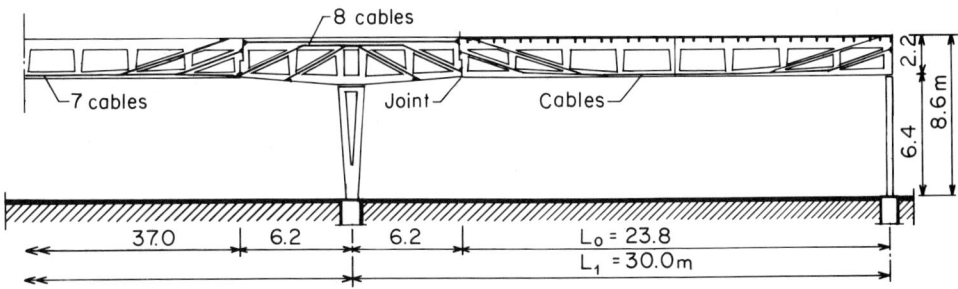

Figure 7-21

8

Arch Action

GENERAL

The true arch action is the equilibrium between the external loads that the arch is subjected to and the internal compressive stresses in the arch, which are uniformly distributed over the cross-sectional area from the crown to the springing of the arch. Such an equilibrium condition is possible when the centerline of the arch coincides with the external-load "force polygon" (line of pressure), as shown in Fig. 8-1.

True arch action is accomplished by one type of stress—compressive stress. Due to this fact, and because the stresses are uniformly distributed over the entire cross-sectional area, the arch action is one of the most economical of the carrying actions.

Arch action is best demonstrated graphically, as in Fig. 8-1. The load to be carried by a symmetrical arch is divided into ten divisions: ΔX = one-tenth of the span L. These loads, $w_1, w_2 \ldots w_{10}$, acting at the center of gravity of the division, are plotted as a force diagram (Fig. 8-1b). Other forces acting on the arch are the horizontal thrust H at the crown and normal forces N at the springings. The shear V_c at the crown is zero because of symmetry. The centerline of the arch is determined graphically by choosing an arbitrary pole $0'$ on the horizontal line drawn from the end of $R_a = \sum_{1}^{5} w_1$ connecting it with a and b (Fig. 8-1). Starting at the centerline of the springing, a parallel is drawn

General

to N_0' until it intersects with load w_1, and so on until the final ray intersects the crown horizontally. The point of intersection c'' of the end lines N_0' and N_6' of the polygon locates the resultant R_a of the loads $w_1 \ldots w_5$. The intersection of R_a and the horizontal line through center of crown c locates the point c'. The line parallel to c' through a locates the correct pole 0 and determines the magnitude of thrust H. The second polygon, with the correct H and carried out in the same manner as the first polygon, which determines the location of the resultant R_a, is the centerline of arch. The rays N_0 to N_{10} are the normal forces in the arch and act at the center of the section.

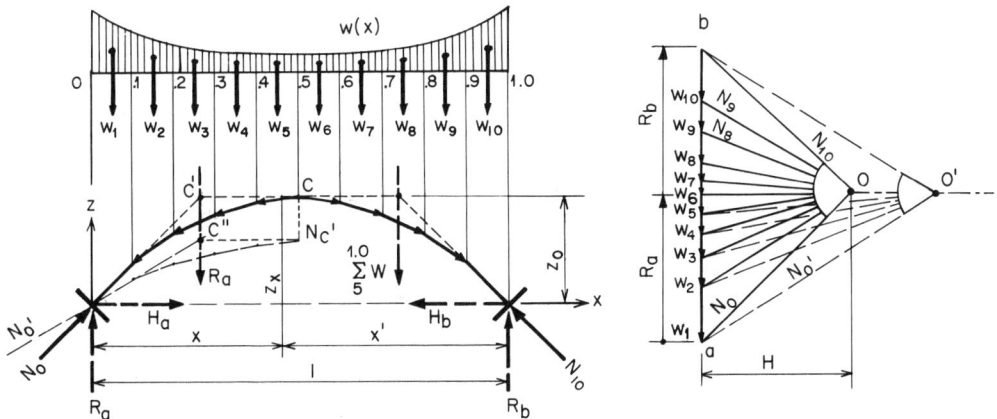

Figure 8-1(a) and (b)

For this centerline of the arch, the equilibrium condition requires that

$$R_a x - w_k(x - a_k) - H z_k = 0 \tag{8-1}$$

for any section of the arch. If this condition is not satisfied, it means that the centerline and pressure line of the arch deviate from each other. This may be caused by differences in actual and assumed loading, by elastic and plastic shortening of the arch, by settlement of supports, and finally by temperature changes. In such a case, the true arch action is lost and equilibrium is established by beam action. The beam action in an arch is represented by the moment

$$M_x = R_0 x - \sum_1^k w_k(x - a_k) - H z_k \tag{8-2}$$

The deviation of the line of pressure from the arch centerline is

$$e_x = \frac{M_x}{H} \tag{8-3}$$

As long as e_x remains within the limits of the kern $\pm d_0/6$ from the centerline, no tensile stresses will occur in the section.

As can be noted in Fig. 8-2, this particular arch curvature is unpleasing, depressive, and not functional because it is flat at the spring, and increasing the useful floor area leads to a high rise. However, it is economical because the curvature is rather close to funicular shape for full dead and half live load; this means that the arch under this load condition is almost entirely in compression and thus the concrete is used most efficiently. From both esthetic and functional points of view, an elliptic or cycloid centerline of the arch would be more acceptable (Fig. 8-2a), but for both of these curvatures the uneconomical beam action ($M = \Delta zH$) is highly involved and practically controls the design. The inefficient use of concrete has to be balanced by increased dimensions, resulting in heaviness. By application of prestressing, the economy of elliptic and cycloid arches can be increased.

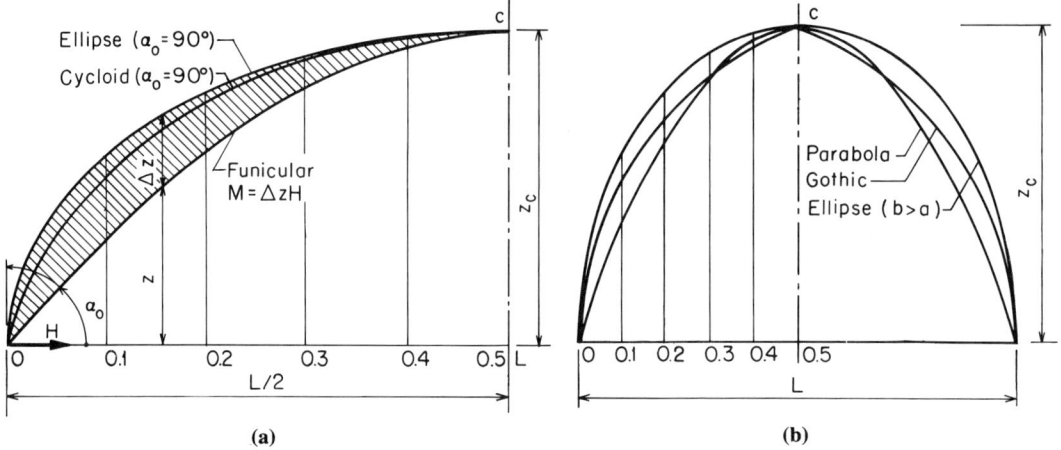

Figure 8-2(a) and (b)

For short spans, high-rise arches (Fig. 8-2b) are visually impressive and economical from a purely structural point of view, but their construction is difficult and expensive. In high-rise arches, normal forces mostly dominate the design because the higher the rise, the larger the compression and the smaller the bending. The moment at crown is zero or rather small. Thus, the proper shape of these arches is structurally similar.

When tensile stresses $-\sigma_0 < +\sigma_B$ (Fig. 8-3) are developed in the section, the factor of safety of an arch is controlled largely by the extent of crack formation. Tensile stresses must be avoided in arches. In view of the fact that the efficiency of beam action M is only one-sixth that of arch action N, the depth of the arch must be increased in accordance with the moments to avoid cracking and to keep the compressive stresses within safe limits. On the other hand, any increase in cross section increases the moment due to temperature change, foundation displacement, and so on, as the third power of the depth (d_0^3). The alternative is to use relatively high normal stresses, as illustrated in Fig. 8-3. However, as a result, the compressive stresses also will increase, the arch will become more flexible, and the danger of buckling of the arch may occur. In addition, in relatively slender arches the redistribution of moment due to plastic flow will be increased.

The economical span-rise ratio (L/z_c) for three-hinged arches can be as high as 10 and for two-hinged arches it can be as high as 8. For fixed arches, it can be as high as 5.

General

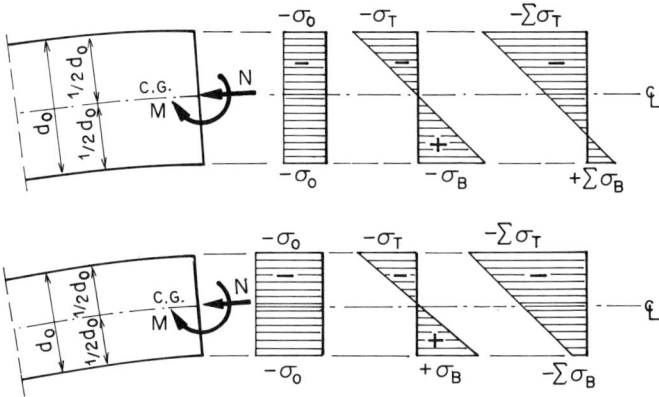

Figure 8-3

The compressive and tensile stresses in the arch are

$$\sigma_x = -\frac{N_x}{A_x} \mp \frac{M_x}{Q_x} \qquad (8\text{-}4)$$

where A_x is the cross-sectional area, N_x is the normal force, and $Q_x = I_x/z_{T,B}$ is the section modulus of the arch (at location x).

There are four types of arches: three-hinged, two-hinged, one-hinged, and hingeless or fixed-end. Each type of arch has its characteristic shape. Selection of the proper type of arch for a particular structural system depends mainly on the perceptual image, the foundation, and the construction method intended. For contemporary reinforced-concrete design, especially in precasting, the most important types are three-hinged and two-hinged arches, which will be discussed in detail.

Theoretically, any arched curvature is possible for any of these arch types. However, for economical reasons, the beam action should be kept to a minimum in an arch. Thus, the curvature selected should not deviate much from the line of pressure. The equations of centerline for circular arch (Fig. 8-4a) and parabolic arch (Fig. 8-4b) are:

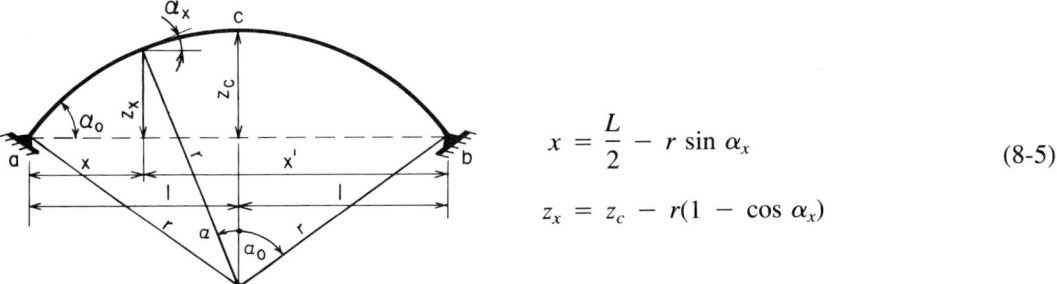

$$x = \frac{L}{2} - r \sin \alpha_x \qquad (8\text{-}5)$$

$$z_x = z_c - r(1 - \cos \alpha_x)$$

Figure 8-4(a)

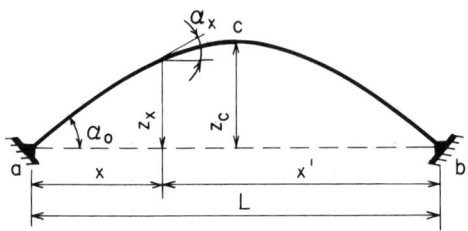

$$z_x = \frac{4z_c}{L^2} xx' \quad x' = L - x$$

$$= 4z_c \, \xi\xi'$$

$$\xi = \frac{x}{L} \quad \xi' = (1 - \xi) \tag{8-6}$$

Figure 8-4(b)

THREE-HINGED ARCHES

Three-hinged arches are statically determinate, which means that the three equations of equilibrium are sufficient to determine the reactions (R), shear (V), and moment (M_0), as for simply supported beams with span L.

For a symmetric three-hinged arch, the thrust (H) is computed from the condition that the moment due to w_D, w_L, or P at the crown (c) is zero. Thus,

$$H = \frac{M_{0c}}{z_c} \tag{8-7}$$

where M_{0c} is the simple beam moment at mid-span and z_c the rise of arch.

The moments due to dead load for parabolic and circular three-hinged arches are computed by Eqs. 8-7 and 8-8, and are illustrated in Fig. 8-5.

For uniform loads, the final moments are determined by Eq. 8-8 and stresses by Eq. 8-4.

The final moments (M_x) and forces (V'_x, N_x) are:

$$M_x = M_{0x} - Hz_x$$

$$V'_x = V_{0x} \cos \alpha_x - H \sin \alpha_x \tag{8-8}$$

$$-N_x = V_{0x} \sin \alpha_x + H \cos \alpha_x$$

For wind load (w_w):

$$R_{0,b} = \mp \frac{w_w z_c^2}{2L}$$

$$H_b = \frac{R_b L}{2z_c}$$

$$H_a = w_w z_c - H_b \tag{8-9}$$

Three-Hinged Arches

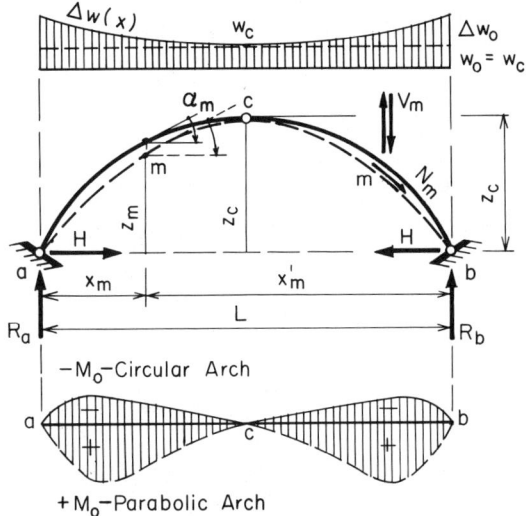

Figure 8-5

$$M_x = R_a x - w_w \frac{z_x^2}{2}$$

$$H_x = -H_a + w_w z_x$$

Deflections of Crown

Temperature change. If the temperature changes, the centerline length of the arch changes, which results in the deflection of the arch. The sagging of crown is

$$\Delta z_c \Delta T = \pm \alpha_t \Delta T \frac{l_1^2 + z_c^2}{z_c} \qquad (8\text{-}10)$$

Due to the deflection, additional moments are introduced in the arch.

$$H_t = \frac{M_{0c}}{z_c \pm \Delta z_c \Delta T}$$

$$M_x = \pm H_t \Delta z_c \Delta T \left(\frac{z_x}{z_c} - \frac{x}{l_1} \right), \quad z' = z \pm \Delta z \qquad (8\text{-}11)$$

Elastic and plastic deformations. The sagging of the crown of a relatively flat and long-span three-hinged arch is considerable over the course of time. The sagging is caused by elastic shortening of the arch length, shrinkage (ε_{sr}), plastic flow (φ), and lateral displacement of the supports (ΔL).

For the time when shrinkage and plastic flow are finished (t = n), the sagging is

$$-\Sigma\Delta z_c = \frac{H_D}{E_0 A} \frac{l_1^2 + z_c^2}{z_c}(1 + \varphi_n) + \varepsilon_{sr}\frac{l_1^2 + z_c^2}{z_c} + \frac{\Delta L l_1}{2z_c} \qquad (8\text{-}12)$$

and because

$$M_{0c} - H_n(z_c - \Sigma\Delta z_c) = 0$$

the thrust for t = n is

$$H_n = \frac{M_{0c}}{z_c - \Sigma\Delta z_c} \qquad (8\text{-}13)$$

Shape of Arch

As can be seen in the moment diagrams (Fig. 8-5), the three-hinged arch has, in almost any typical loading condition, maximum moments approximately at the quarter points, which decrease toward the crown and spring. The presence of a moment in a section reduces the efficient use of material considerably. At the spring and crown, where the moment is zero and only the normal force N is acting at the center of gravity, the material is used 100 percent, but at the quarter point only up to 50 percent is used. To keep the stresses within safe limits, the cross section must be increased to balance the inefficient use of the material. The proper shape of a three-hinged arch is shown in Fig. 8-6.

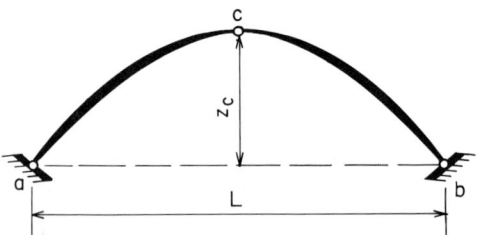

Figure 8-6

Example

$$M_a = M_b = M_c = 0$$

$$L = 50 \text{ m}$$

$$z_c = \frac{L}{4} = \frac{50}{4} = 12.5 \text{ m}$$

$$w = 1.0 \text{ t/m}$$

$$M_{c0} = \frac{1}{8}wL^2 = \frac{1}{8}1.0 \cdot 50^2$$

$$= 313 \text{ tm}$$

Two-Hinged Arches

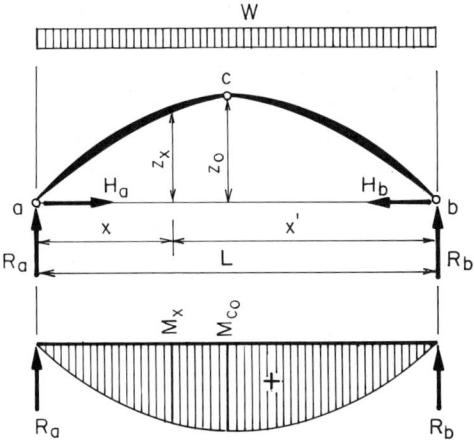

Figure 8-7

$$H = \frac{M_{c0}}{z_c} = \frac{313}{12.5} = 25 \text{ t}$$

$$z_x = 4z_c\xi\xi' \quad - \text{ parabolic arch}$$

$$\xi = \frac{x}{L} \qquad \xi' = \frac{x'}{L}$$

Moments and reactions: Parabola is the pressure line for uniform loading (moments—zero)

$$x = 17.00 \text{ m}, \, x' = 33.00 \text{ m}: \qquad \xi = 0.34, \, \xi' = 0.66$$

$$M_x = 4M_{c0}\xi\xi' = 4 \cdot 313 \cdot 0.34 \cdot 0.66 = 282 \text{ tm}$$

$$z_x = 4 \cdot z_c\xi\xi' = 4 \cdot 12.50 \cdot 0.34 \cdot 0.66 = 11.25 \text{ m}$$

$$\Delta M = 313 - 25 \cdot 12.5 = 0$$

$$R_a = R_b = \frac{wL}{2} = \frac{1.0 \cdot 50}{2} = 25 \text{ t}$$

TWO-HINGED ARCHES

A two-hinged arch is one-time statically indeterminate. The redundant is the thrust H, which can be computed from the deformations of the arch. A simple beam serves as the principal system. The condition for computing is that the horizontal displacement of the hinges be zero.*

*A. E. Komendant, *Contemporary Concrete Structures* (New York: McGraw-Hill Book Company, 1972).

The redundant thrust H depends on the deformation of the arch subjected to various loadings. In a principal system the thrust H is zero; therefore, the loading causes lateral displacement δ_0 of the arch. After δ_0 is computed, the force required to eliminate the displacement δ_0 is determined, as illustrated in Fig. 8-8.

The maximum moment of a two-hinged arch occurs at the crown; therefore, the most suitable shape of the arch is described by the formula

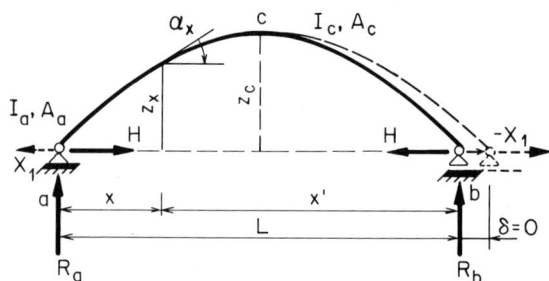

Figure 8-8

$$\frac{I_c}{I_x \cdot \cos \alpha_x} = 1 - (1 - n)(1 - 2\xi)^2 \tag{8-14}$$

$$\xi = \frac{x}{L}$$

$$n = \frac{I_c}{I_a \cos \alpha_0}$$

$$n = 1: \quad I = \text{constant}$$

$$z_x = 4z_c \xi \xi'$$

For example,

$$\delta_{01} - X_1 \delta_{11} = 0 \qquad X_1 = \frac{\delta_{01}}{\delta_{11}} = H \tag{8-15}$$

The displacement δ_{11} due to $X_1 = -1$ for parabolic arch is

$$\delta_{11} = \frac{8}{15} \frac{6+n}{7} z_c^2 L(1 + \nu)$$

$$\nu = \frac{15}{8} \frac{7}{6+n} \frac{1}{z_c^2} \frac{I_c}{A_c} \tag{8-16}$$

Two-Hinged Arches

The reactions (R), thrust (H), and moments (M) for a parabolic symmetrical two-hinged arch for common loadings are given in Fig. 8-9.

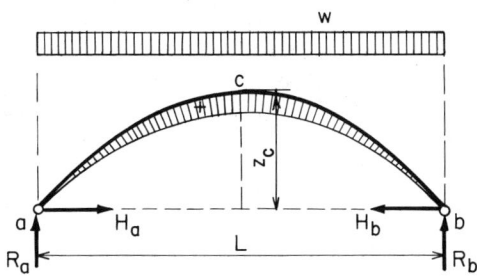

Figure 8-9(a)

$$R_a = R_b = \frac{wL}{2}$$

$$H_a = H_b = \frac{wL^2}{8z_c(1 + \nu)}$$

$$M_c = \frac{wL^2}{8} \frac{\nu}{1 + \nu}$$

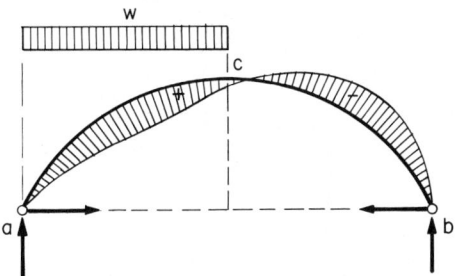

Figure 8-9(b)

$$R_a = \frac{3}{8}wL \qquad R_b = \frac{1}{8}wL$$

$$H_a = H_b = \frac{wL^2}{16z_c(1 + \nu)}$$

$$M_c = \frac{wL^2}{16} \frac{\nu}{1 + \nu}$$

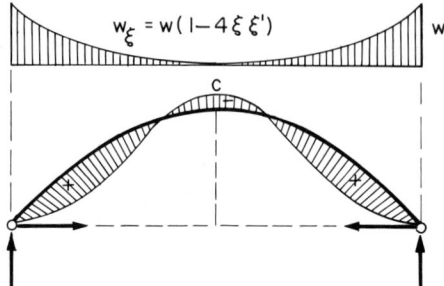

Figure 8-9(c)

$$R_a = R_b = \frac{wL}{6}$$

$$H_a = H_b = \frac{wL^2}{36z_c(1+\nu)} \frac{5+n}{6+n}$$

$$M_c = M_{0c} - H_a z_c$$

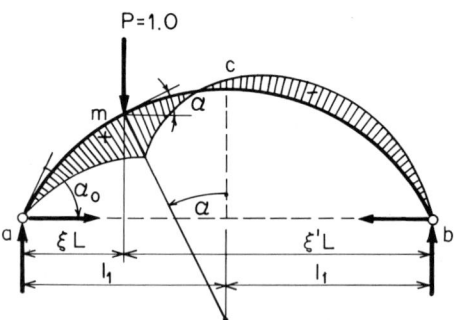

Figure 8-9(d)

$$R_a = \xi'P \qquad R_b = \xi P$$

$$H_a = H_b = \frac{LP}{8z_c(1+\nu)} \frac{7\xi\xi'}{6+n}$$

$$[5(1 + \xi\xi') + (n-1)$$

$$(1 + \xi\xi' - 8\xi^2\xi'^2)]$$

Two-Hinged Arches

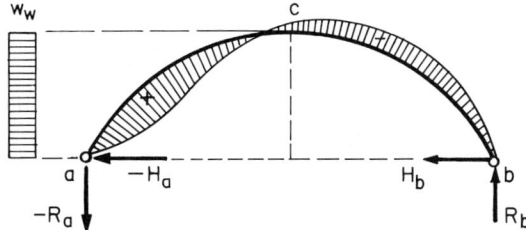

Figure 8-9(e)

$$-R_a = R_b = \frac{w_w z_c^2}{2L}$$

$$H_a = -0.714 w_w z_c$$

$$H_b = w_w z_c - H_a = 0.286 w_w z_c$$

Temperature change ΔT and lateral displacement of supports ΔL:

$$R_a = R_b = 0$$

$$H_{a,b} = \frac{15 E_0 I_c (\alpha_t \Delta T L - \Delta L)}{8 z_c^2 L (1 + \nu)} \qquad (8\text{-}17)$$

$$M_x = H z_x$$

The deflections of the crown due to elastic shortening (ε_e) and plastic flow (ε_{pl}) of the arch cancel each other out. The shortening of the arch due to shrinkage (ε_{sr}) and horizontal displacement of supports (ΔL) is considerably reduced by plastic flow. Therefore, the change of thrust ($\Delta H_{\Delta T}$) can be assumed to be zero.

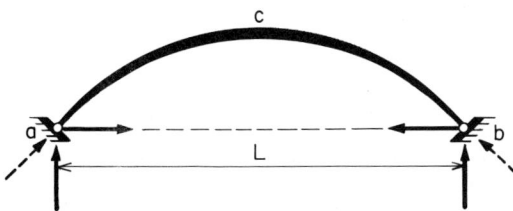

Figure 8-10

The moment reaches its maximum at the crown and gradually decreases toward the springs. The proper shape of a two-hinged arch is shown in Fig. 8-10.

FIXED ARCHES

Fixed arches are three times statically indeterminate. The simply supported curved beam most conveniently serves as the principal system. The redundants are the thrust (H) and fixed-end moments (M_a, M_b). For symmetric loadings, $M_a = M_b$ (Fig. 8-11).

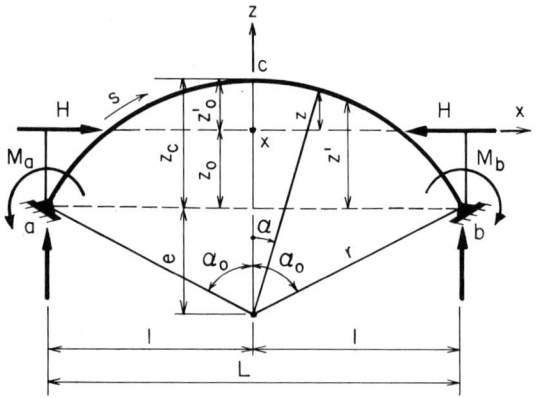

Figure 8-11

For symmetric parabolic arch having a form

$$\frac{I_c}{I \cos \alpha}$$

$$z_0 = \frac{2}{3} z_c \qquad (8\text{-}18)$$

$$\nu = \frac{45}{4} \frac{I_c}{A_c z_c^2}$$

For typical loadings, the reactions, thrust, and moments are given by the following equations (Fig. 8-12).

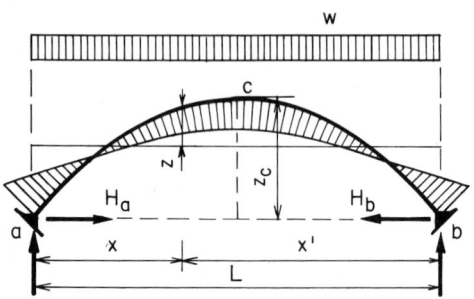

Figure 8-12(a)

Fixed Arches

$$R_a = R_b = \frac{wL}{2}$$

$$H_a = H_b = \frac{wL^2}{8z_c}\frac{1}{1+\nu}$$

$$M_a = M_b = -\frac{wL^2}{12}\frac{\nu}{1+\nu}$$

$$M_c = +\frac{wL^2}{24}\frac{\nu}{1+\nu}$$

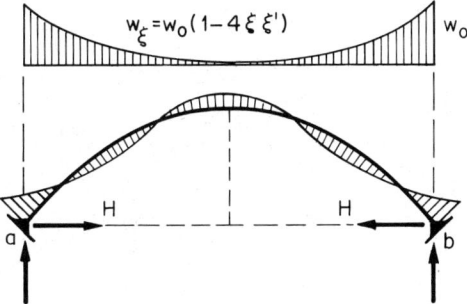

Figure 8-12(b)

$$R_a = R_b = \frac{w_0 L}{6}$$

$$H = \frac{w_0 L^2}{56 z_c}\frac{1}{1+\nu}$$

$$M_a = M_b = -\frac{w_0 L^2}{420}\frac{7\nu + 2}{1+\nu}$$

$$M_c = -\frac{w_0 L^2}{1680}\frac{3 - 7\nu}{1+\nu}$$

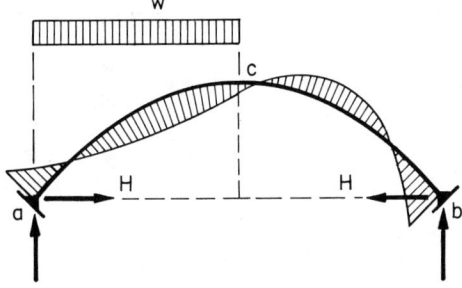

Figure 8-12(c)

$$R_a = \frac{13}{32} wL \qquad R_b = \frac{3}{32} wL$$

$$H = \frac{wL^2}{16z_c} \frac{1}{1+\nu}$$

$$M_a = -\frac{wL^2}{192} \frac{3+11\nu}{1+\nu}$$

$$M_b = \frac{wL^2}{192} \frac{3-5\nu}{1+\nu}$$

$$M_c = \frac{wL^2}{48} \frac{\nu}{1+\nu}$$

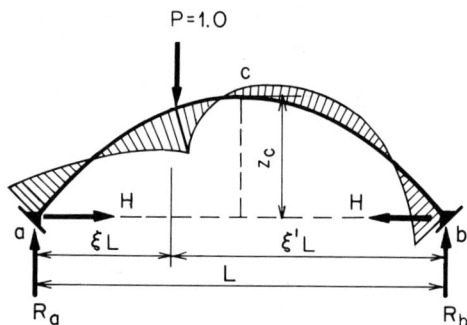

Figure 8-12(d)

$$R_a = \xi'^2(1 + 2\xi)$$

$$R_b = \xi^2(1 + 2\xi')$$

$$H = \frac{15L}{4z_c} \frac{\xi^2 \xi'^2}{(1+\nu)}$$

$$M_a = \xi\xi'^2 LP\left[\frac{5}{2(1+\nu)} \xi - 1\right]$$

$$M_b = \xi^2\xi' LP\left[\frac{5}{2(1+\nu)} \xi' - 1\right]$$

$$M_c = \xi^2 \frac{LP}{2}\left[1 - \frac{5}{2(1+\nu)} \xi'^2\right]$$

Fixed Arches

Temperature change:

$$\delta_{0t} - X\delta_{11} = 0$$

$$X = \frac{\delta_0 t}{\delta_{11}} = H$$

$$\pm \delta_{0t} = \alpha_t \Delta TL$$

$$EI_c \delta_{11} = \frac{4}{45} Lz_c^2 (1 + \nu)$$

$$X = \pm \frac{EI_c \alpha_t \Delta TL}{\frac{4}{45} Lz_c^2 (1 + \nu)} = H$$

$$M_a = M_b = \pm Hz_0$$

9

Space Structures

GENERAL

The particular feature of a shell is the transmission of external loads (w) into the supports mainly by direct stresses with relatively small flexural stresses. As the material is most efficiently used when only direct stresses are involved in carrying action, relatively large areas can be bridged economically without any intermediate supports.

The axis of the differential element is denoted by x, y, and z, where x is tangent to the horizontal curve or direction of a straight section, y is tangent to meridian or transversal curvature, and z is normal to the element. The shells are statically determinate when the shell's edges at support are free to rotate and displacements are allowed to occur. In this condition, the external loads (w) are transmitted to supports by membrane (direct) stresses, which resultants (N_x, N_α, N_β, $N_{\alpha\beta}$) can be computed by the three equations of equilibrium.

Equilibrium requires that all forces in the x-, y-, and z-directions be zero. Thus

$$\Sigma X = 0, \Sigma Y = 0, \Sigma Z = 0$$

Loading

The external loads, which the shells are commonly subjected to, are the dead load w_D and live load w_L. For symmetric loading $w_x = 0$, w_y and w_z are functions of α and independent of x (Fig. 9-1).

Domes

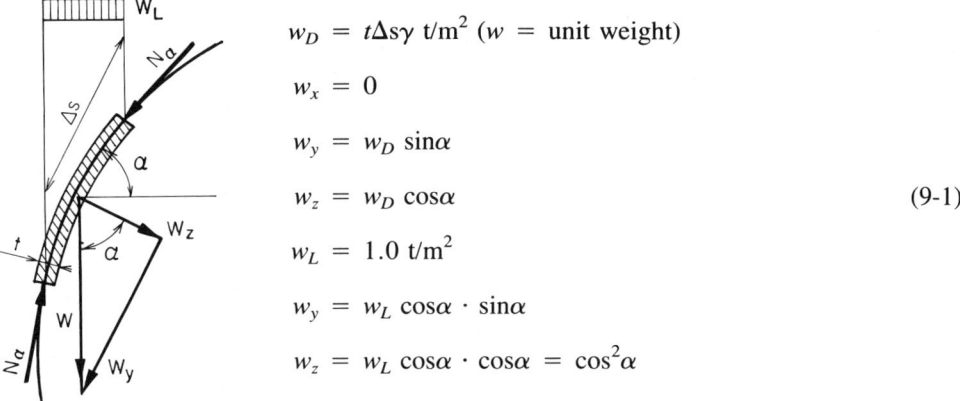

$$w_D = t\Delta s\gamma \text{ t/m}^2 \ (w = \text{unit weight})$$
$$w_x = 0$$
$$w_y = w_D \sin\alpha$$
$$w_z = w_D \cos\alpha$$
$$w_L = 1.0 \text{ t/m}^2$$
$$w_y = w_L \cos\alpha \cdot \sin\alpha$$
$$w_z = w_L \cos\alpha \cdot \cos\alpha = \cos^2\alpha$$

(9-1)

Figure 9-1

DOMES

For domes or rotation symmetrical spherical shells, the meridian section can be a circle, ellipse, cycloid, cone, or any other symmetric curvature. The horizontal sections are always circles.

With the rotation symmetrical loading $w_x = 0$ and shear $N_{\alpha\beta} = 0$, the N_α-value can be most simply determined by Guldin's rule (Fig. 9-2).

$$r_\alpha = R \sin\alpha$$
$$N_\alpha = w_{z\alpha} R \qquad N_\beta = w_{z\beta} R$$
$$H_\alpha = N_\alpha \cos\alpha \qquad T_o = w_z r_o$$

Membrane Forces

Equilibrium requires: $w_{z\alpha} + w_{z\beta} = w_z \qquad \dfrac{N_\alpha}{R} + \dfrac{N_\beta}{R} = w_z$

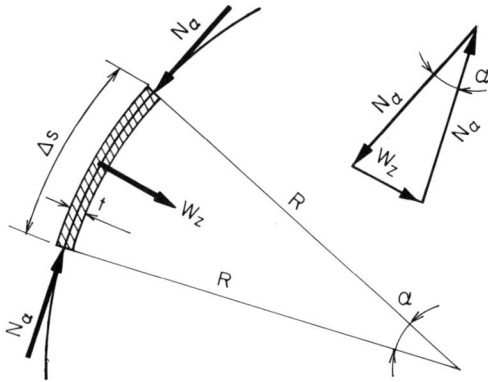

Figure 9-2(a)

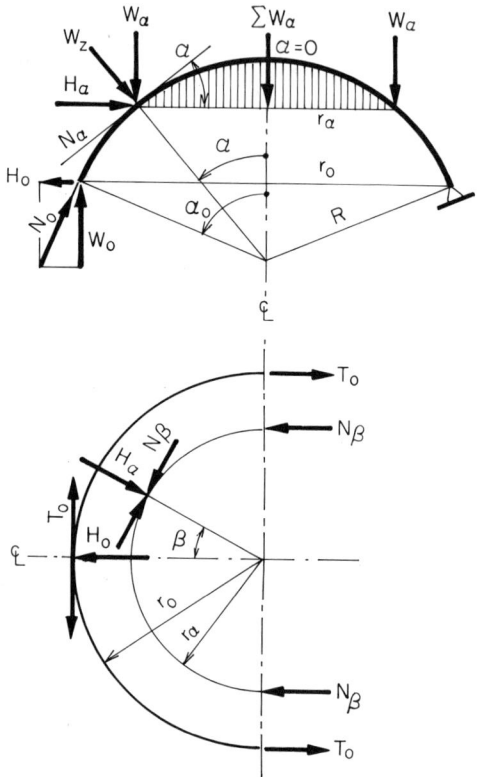

Figure 9-2(b)

$$A_\alpha = 2\pi R^2(1 - \cos \alpha)$$

$$\Sigma w_\alpha = A_\alpha w_D = 2\pi R^2 w_D(1 - \cos \alpha)$$

$$w_\alpha = \frac{\Sigma w_\alpha}{2\pi r_\alpha} = \frac{Rw_D(1 - \cos \alpha)}{\sin \alpha} \tag{9-2}$$

$$N\alpha = -\frac{Rw_D(1 - \cos \alpha)}{\sin^2 \alpha}$$

$$N_\alpha + N_\beta = Rw_z$$

$$N_\beta = Rw_z - N_\alpha$$

Dead load.

$$w_y = w_D \sin \alpha, \; w_z = w_D \cos \alpha, \; w_x = 0$$

$$N_\alpha = -\frac{Rw_D(1 - \cos \alpha)}{\sin^2 \alpha} \tag{9-3}$$

Domes

$$= -\frac{Rw_D}{1 + \cos \alpha}$$

$$N_\beta = Rw_z - N_\alpha$$

$$= \frac{Rw_D}{1 + \cos \alpha}(1 - \cos \alpha - \cos^2 \alpha)$$

$$\alpha = 0: \qquad \cos \alpha = 1$$

$$N_\alpha = N_\beta = -\frac{Rw_D}{2}$$

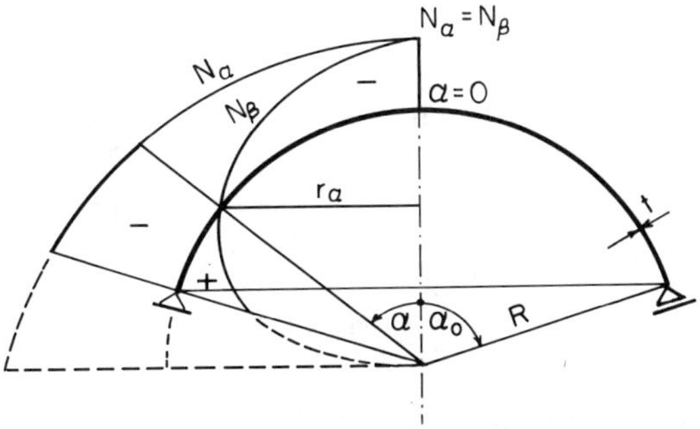

Figure 9-3

Live load.

$$w_y = w_L \sin \alpha \cos \alpha \qquad w_z = w_L \cos^2 \alpha \qquad w_x = 0$$

$$\Sigma w_\alpha = \pi r_\alpha^2 w_L = \pi R^2 \sin^2 \alpha \, w_L \tag{9-4}$$

$$N_\alpha = -\frac{\Sigma w_\alpha}{2\pi R \sin^2 \alpha} = \frac{\pi R^2 \sin^2 \alpha \, w_L}{2\pi R \sin^2 \alpha} = -\frac{Rw_L}{2}$$

$$N_\beta = -\frac{Rw_L}{2} \cos 2\alpha$$

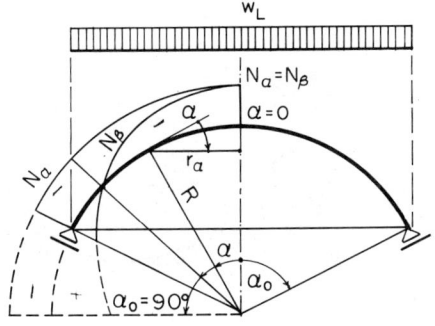

Figure 9-4

$$\alpha = 0: \quad \cos\alpha = 1$$

$$N_\alpha = N_\beta = -\frac{Rw_L}{2}.$$

As can be seen from the diagrams, the N_α is a compressive force from $\alpha = 0°$ to $\alpha = 90°$. The N_β is a compressive force for dead load until $\alpha \leq 51°50'$ and for live load $\alpha \leq 45°$. For greater α-values, N_β becomes a tensile force.

Wind load. For wind load: $w_x = 0$, $w_y = 0$, $w_z = w_w \sin\alpha \cos\beta$

$$W_z = \frac{\pi R^2}{3} w_w (2 - 3\cos\alpha + \cos^3\alpha)$$

$$N_\alpha = -\frac{R \cos\alpha}{3 \sin^3\alpha} w_w (2 - 3\cos\alpha + \cos^3\alpha) \cos\beta$$

$$N_\beta = \frac{R}{3 \sin^3\alpha} w_w (2\cos\alpha - 3\sin^2\alpha - 2\cos^4\alpha) \cos\beta \qquad (9\text{-}5)$$

$$N_{\alpha\beta} = -\frac{R}{3 \sin^3\alpha} w_w (2 - 3\cos\alpha + \cos^3\alpha) \sin\beta$$

Reactions R and H for all loadings can be simply computed from equilibrium conditions.

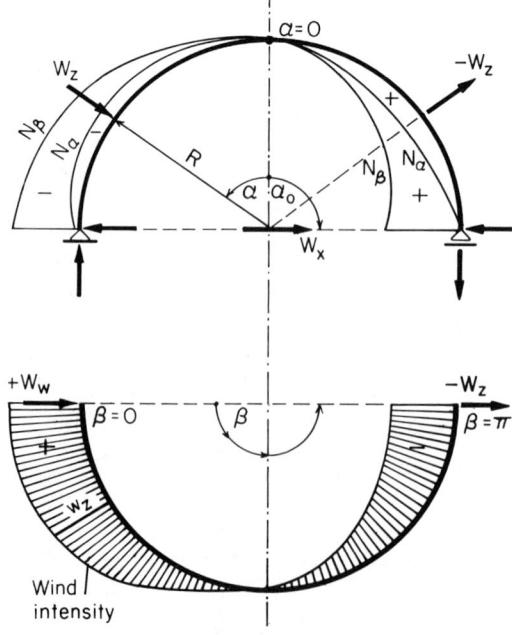

Figure 9-5

Domes

The wind intensity and N_α and N_β diagrams are shown in Fig. 9-5.

Marginal Member

In order to induce the vertical reactions (R), tangent to the shell at $\alpha = \alpha_0$, a horizontal force in the magnitude of $H = N_{\alpha 0} \cos\alpha_0$ is required (Fig. 9-6).

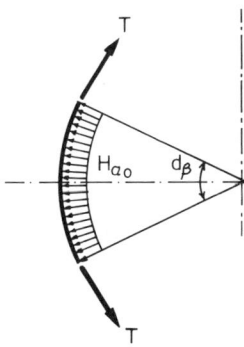

Figure 9-6

For dead and live loads, the H-forces are:

$$H_D = -Rw_D \frac{\cos\alpha_0}{1 + \cos\alpha_0}$$

$$H_L = -Rw_L \frac{\cos\alpha_0}{2} \qquad (9\text{-}6)$$

$$T = H_{\alpha_0} r_0$$

Deformations

As a rule, the horizontal displacement Δr and the rotation of meridian tangent δ of the shell are adequate to describe the deformations of the spherical shell. They can be simply expressed by strain or directly by $N_\alpha N_\beta$ values. Assuming Poisson's ratio $\nu = 0.167$, they are for dead and live loads, and temperature change.

$$\Delta r_\alpha = \frac{R^2 w_D}{E\,t} \left(\frac{1 + \nu}{1 + \cos\alpha} - \cos\alpha \right) \sin\alpha$$

$$\vartheta_\alpha = \frac{R w_D}{E\,t} (2 + \nu) \sin\alpha \qquad (9\text{-}7)$$

$$\Delta r_\alpha = \frac{R^2 w_L}{E\,t} \left(\frac{1 + \nu}{2} - \cos^2\alpha \right) \sin\alpha$$

$$\vartheta_\alpha = -\frac{Rw_L}{Et}(3+\nu)\sin\alpha \cdot \cos\alpha$$

$$\Delta r_\alpha = \alpha_t \, \Delta TR \sin\alpha$$

$$\vartheta_\alpha = 0$$

The bending moments in the dome are of secondary order, maximum at the boundary (α_0), and decrease very rapidly toward the crown ($\alpha = 0$).

Example

Dead load:

$$R = 50m \qquad \alpha = 36°, \sin\alpha = 0.588, \cos\alpha = 0.809$$

$$\alpha_0 = 70° \qquad w_D = 0.12 \cdot 2.400 \cong 0.29 \text{ t/m}^2$$

$$N_\alpha = -\frac{Rw_D}{1+\cos\alpha} = \frac{50 \cdot 0.29}{1+0.809} = -8.0 \text{ t/m}$$

$$N_\beta = \frac{Rw_0}{1+\cos\alpha}(1-\cos\alpha - \cos^2\alpha)$$

$$= 8.0\,(1 - 0.809 - 0.809^2) = -3.70 \text{ t/m}$$

$\alpha = 0$:

$$N_\alpha = N_\beta = -\frac{Rw_D}{2} = -\frac{50 \cdot 0.29}{2} = -7.25 \text{ t/m}$$

$$\alpha_0 = 70°: \cos\alpha_0 = 0.342, \sin\alpha_0 = 0.940$$

$$N_{\alpha_0} = \frac{50 \cdot 0.29}{1+0.342} \cong -11.0 \text{ t/m}$$

$$N_\beta = 11.0\,(1 - 0.342 \cdot 0.342^2) = 5.9 \text{ t/m}$$

Edge member:

$$N_{\alpha_0} = -11.0 \text{ t/m}, \; H_{\alpha_0} = N_{\alpha_0}\cos\alpha = -11.0 \cdot 0.342 = -3.76 \text{ t/m}$$

$$T_{\alpha_0} = r_{\alpha_0} H_{\alpha_0} = R\sin\alpha_0 \, H_{\alpha_0} = 50 \cdot 0.940 \cdot 3.76 = 177 \text{ t}$$

Live load:

$$\alpha = 36°:$$

$$w_z = w_L \cos^2\alpha \qquad w_L = 0.20 \text{ t/m}^2$$

$$w_y = w_z \cos\alpha \cdot \sin\alpha$$

$$\Sigma w_\alpha = \pi r_\alpha^2 \, w_L, \; r = R\sin\alpha$$

Domes

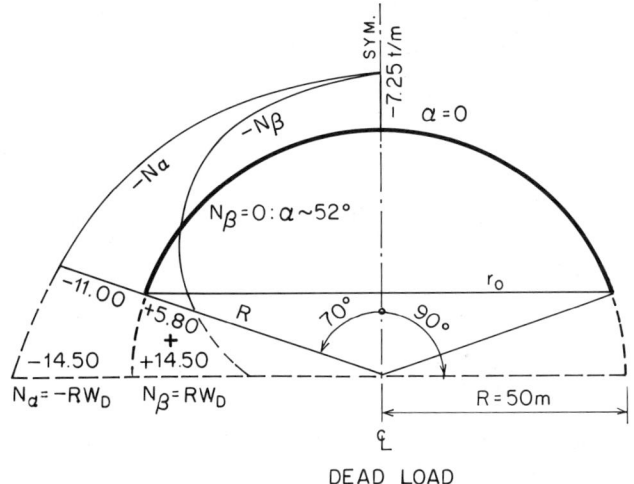

DEAD LOAD

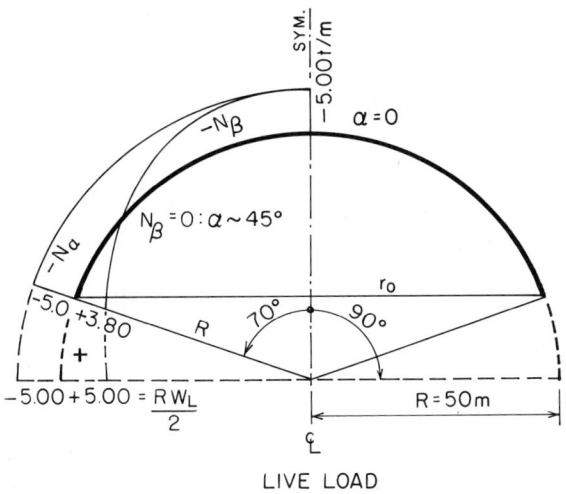

LIVE LOAD

Figure 9-7

$$N_\alpha = -\frac{\Sigma w_\alpha}{2\pi r_\alpha \sin\alpha} = -\frac{R \cdot w_L}{2} = -\frac{50 \cdot 0.20}{2} = -5.0 \text{ t/m} = \text{const.}$$

$$N_\beta = N_\alpha + Rw_L \cos^2\alpha = -5.0 + 50 \cdot 0.20 \cdot 0.342^2 = -3.83 \text{ t/m}$$

$\alpha_0 = 70°$:

$$N_{\alpha_0} = -5.0 \text{ t/m} \qquad H_{\alpha_0} = \cos\alpha_0 = -5.0 \cdot 0.342 \cong -1.70 \text{ t/m}$$

Edge member:

$$T_{\alpha_0} = R \sin\alpha_0 \, H_{\alpha_0} = 50 \cdot 0.94 \cdot 1.70 = 80 \text{ t}$$

$$\sum_D^L T_{\alpha_0} = 177 + 80 = 257 \text{ t}$$

The normal forces for dead and live loads are shown in Fig. 9-7.

FOLDED PLATES

Simplified Theory

The transversal section of a folded plate is composed of slabs forming a polygon. The folded plate or polygonlinear shell acts as a special type of barrels. If the ratio z_c/B is relatively large, the yielding of ridges very often is negligible. However, if the ratio is small, the deflection of ridges is larger and more rigorous analysis is required.* In the first case, the plates in the transversal direction act as continuous slabs on rigid supports and in the second case as continuous slabs on yielding supports. The plates in longitudinal direction act as simple or continuous beams supported on rigid frames or diaphragms (Fig. 9-8).

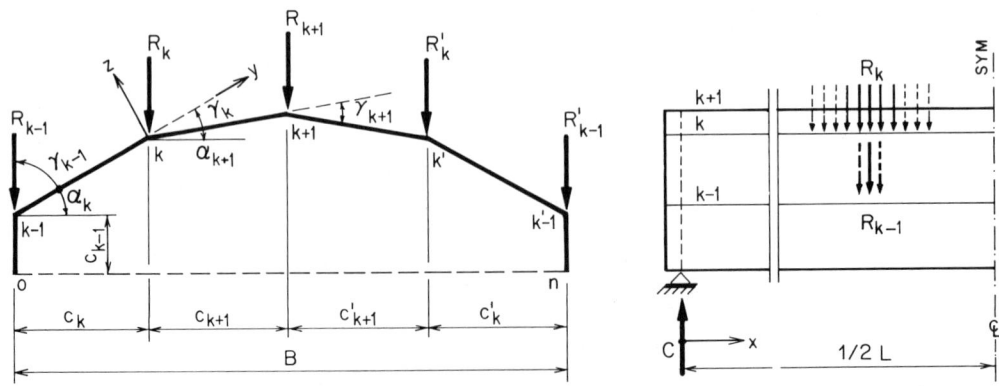

Figure 9-8

*A. E. Komendant, D.E., *Contemporary Concrete Structures* (New York: McGraw-Hill Book Company, 1972).

Folded Plates

Panel loads. The slab reactions R_k produce uniform line loads from $x = 0$ to $x = L$ for ridges, which are transmitted along the polygon panels in the x-direction into end supports C, C'. The panel loads P_k are (Fig. 9-9).

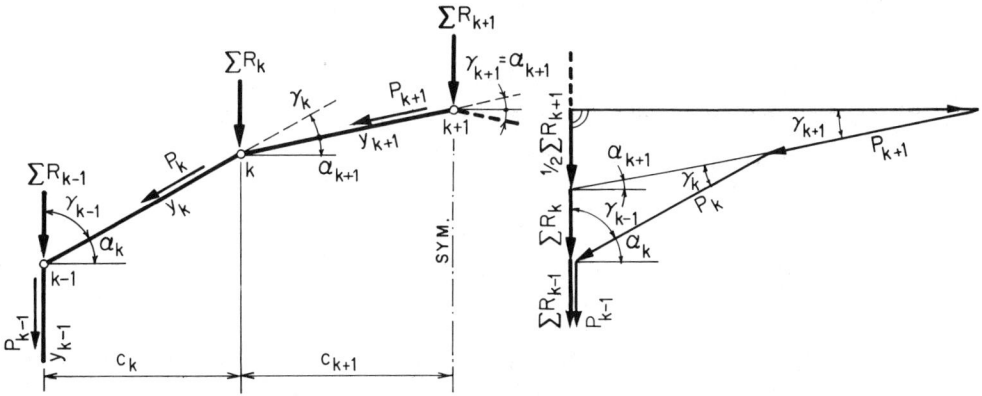

Figure 9-9

$$P_{k+1} = \frac{1}{2} \sum R_{k+1} \frac{1}{\sin \gamma_{k+1}} - \sum R_k \frac{\cos \alpha_k}{\sin \gamma_k}$$

$$P_k = \sum R_k \frac{\cos \alpha_{k+1}}{\sin \gamma_k} - \sum R_{k-1} \frac{\cos \alpha_{k-1}}{\sin \gamma_{k-1}} \qquad (9\text{-}8)$$

$$P_{k-1} = \sum R_{k-1}$$

Individual panels in a polygon subjected to panel loads P_k differ from ordinary beams only by boundary condition of the top and bottom fibers. At the common boundary (ridge k) the adjacent panels must meet the condition

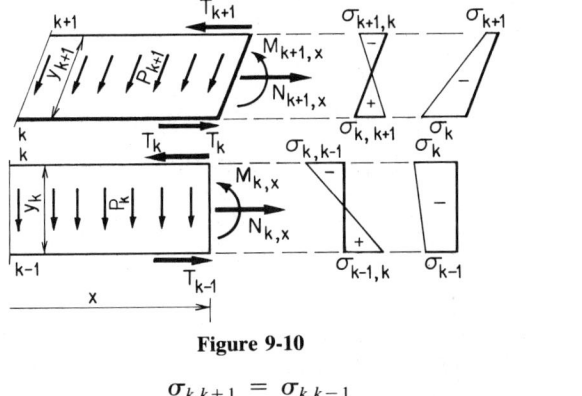

Figure 9-10

$$\sigma_{k,k+1} = \sigma_{k,k-1} \qquad (9\text{-}9)$$

$$\frac{d\sigma_{k,k+1}}{dx} = \frac{d_{k,k-1}}{dx}$$

Shear forces. This condition can be satisfied only by shear force T_k. Equilibrium requires that (Fig. 9-10)

$$\Sigma N_x = 0: \quad N_k = T_k - T_{k-1}$$

$$\Sigma M_x = 0: \quad M_k = M_{k0} - \frac{1}{2} y_k (T_k + T_{k-1}) \qquad (9\text{-}10)$$

The beam moment for uniform loading P_k is

$$M_{k0,\max} = \frac{P_k L^2}{8} \qquad M_{k0x} = 4 M_{k0,\max} \xi \xi' \qquad (9\text{-}11)$$

Fiber stresses. The requirement that the stresses $\sigma_{k,x}$ in adjacent panels along their common ridge must be equal allows the determination of the shear force T_k. Considering that the cross-sectional area (A) and section modulus (Q_k) are

$$A_k = t\, y_k \qquad Q_k = \frac{t\, y_k^2}{6} \qquad \text{and} \quad \frac{M_{k,0}}{Q_k} = \sigma_{k,0}$$

The fiber stresses $\sigma_{k,k+1}$ and $\sigma_{k,k-1}$ of the panels at ridge k are:

$$\sigma_{k,k+1} = \frac{N_{k+1}}{A_{k+1}} + \frac{M_{k+1,0} - 1/2(T_k + T_{k+1})}{Q_{k+1}} \qquad (N_{k+1} = T_{k+1} - T_k)$$

$$= \sigma_{k+1,0} - \frac{4}{A_{k+1}} T_k - \frac{2}{A_{k+1}} T_{k+1} \qquad (9\text{-}12)$$

$$\sigma_{k,k-1} = -\frac{N_k}{A_k} - \frac{M_{k,0} - 1/2(T_k + T_{k-1})}{Q_k} \qquad (N_k = T_k\, T_{k-1})$$

$$= -\sigma_{k,0} + \frac{4}{A_k} T_k + \frac{2}{A_k} T_{k-1}$$

Equating these equations, we obtain

$$\frac{\sigma_{k+1,0} + \sigma_{k,0}}{2} = \frac{1}{A_k} T_{k-1} + 2 \left(\frac{1}{A_k} + \frac{1}{A_{k+1}} \right) T_k + \frac{1}{A_{k+1}} T_{k+1} \qquad (9\text{-}13)$$

Such a linear equation can be written for any ridge and the solution of these simultaneous equations delivers the shear T_{k-1}, T_k, T_{k+1} ($k = 1, 2, \ldots, n$). Then the σ_k-stresses can be computed from the Eq. 9-12.

The panel (slab) ends at y-direction can be assumed fixed at the ridges. The slab loadings are given by Eqs. 9-8 and 9-9 (Fig. 9-9).

Curvilinear Cylindrical Shells (Barrels)

The folded plate type shown in Fig. 9-11 can be computed in the same way as simple beams.

The sectional coefficients for a section c can be computed by Eqs. 1-10 and 1-12. The moment and stresses due to total dead load (w_D) and live load (w_L), to which the section is subjected, are for simple span:

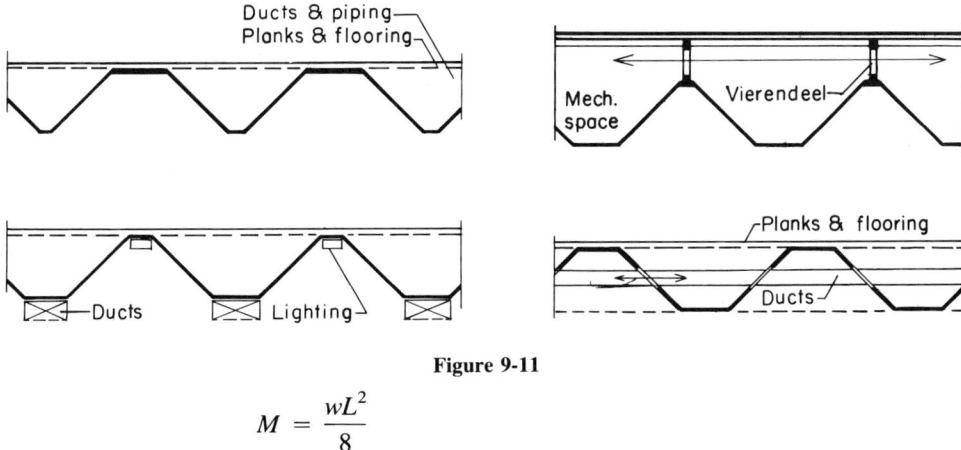

Figure 9-11

$$M = \frac{wL^2}{8}$$

$$\sigma_{T,B} = \mp \frac{M}{I_0} z_{T,B}$$

$$v_{max} = \frac{VQ}{I_0 b_0}$$

CURVILINEAR CYLINDRICAL SHELLS (BARRELS)

Membrane Theory

The transversal curvature of barrels can be circular, elliptical, cycloidal, or of any curvature in which the centerline remains higher than the pressure line of the loading to which the shell is subjected (Fig. 9-12).

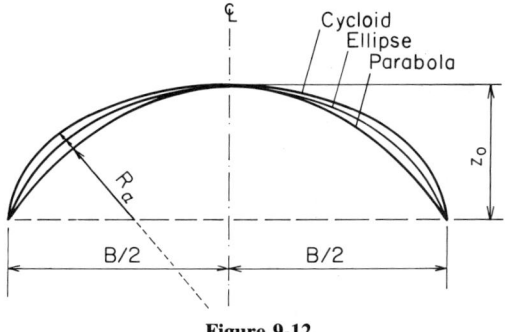

Figure 9-12

As the pressure line for uniform horizontal loading is a parabola, no shell action occurs and the load is transmitted directly by arch action into the springs. The longitudinal or horizontal sections are rectangular. The barrels are supported at the ends or at intermediate points by arches, arched frames, or diaphragms. The barrels are relatively rigid in the vertical direction but rather flexible horizontally. They act and can be analyzed as a simple or continuous beam, using a curved cross section (Fig. 9-13).

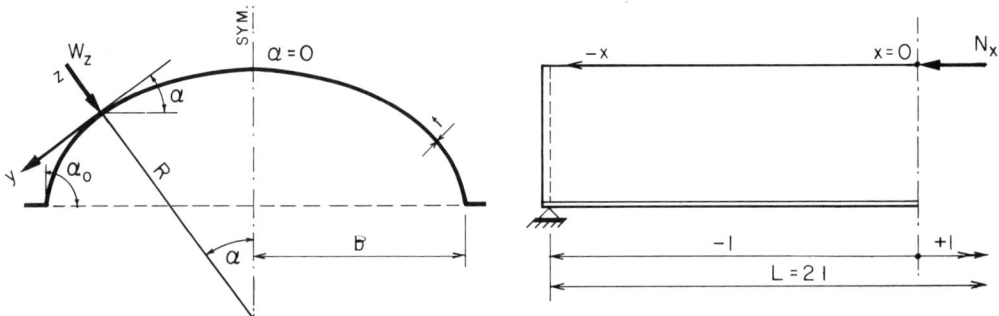

Figure 9-13

The main differences in structural behavior between the beam and the cylindrical shell lie in the normal stress distribution over the cross section in the vertical direction and in the distinctive action of the shell in the transversal direction. These two basic differences are explained by the fact that the cross section in the transversal direction is rigid in the case of a beam but rather elastic in the case of a shell. However, there are cases in which these deviations are of secondary order and can be disregarded. In these cases, the reactions and internal forces in the shell due to uniform external loads can be determined with sufficient accuracy by the three equations of equilibrium or by membrane theory.

The basic requirements for applicability of membrane theory are:

1. The curvature of the cross section of the shell (meridian curvature) should have vertical or almost vertical tangents at its longitudinal edge ($\alpha_0 = 90°$).
2. The boundary conditions must be satisfied.

If these two basic requirements cannot be met—for architectural, practical, or any other reasons—relatively large flexural stresses are developed in the shell that very often penetrate from one longitudinal boundary to the other and entirely control the design. In such cases, use must be made of the bending theory of shells.

To explain the shell action, the shell element illustrated in Fig. 9-14 will be considered. Using rectangular coordinates, the equilibrium between external loads and internal forces requires that

$$\Sigma X = 0: \quad \frac{\partial N_x}{\partial x} + \frac{\partial N_{x\alpha}}{R \partial x} + w_x = 0$$

Curvilinear Cylindrical Shells (Barrels)

$$\Sigma Y = 0: \qquad \frac{\partial N_{x\alpha}}{\partial \alpha} + \frac{\partial N_\alpha}{R \partial \alpha} + w_y = 0 \tag{9-14}$$

$$\Sigma Z = 0: \qquad N_\alpha + R w_z = 0$$

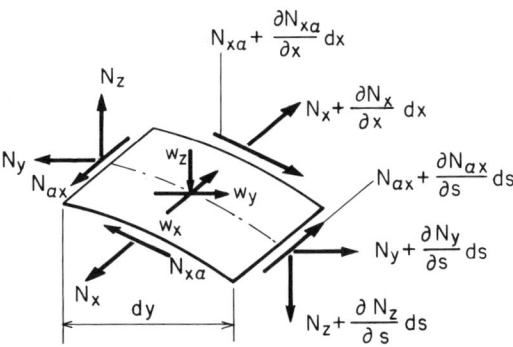

Figure 9-14

Membrane Forces

The shear force $N_{x\alpha}$ and normal force N_x can be obtained by direct integration with respect to x, and N_α is directly a function of w_z. Thus,

$$N_{x\alpha} = -w_y^* x, \qquad w_y^* = w_y + \frac{\partial N_\alpha}{\partial \alpha}$$

$$N_x = -\frac{1}{R} \frac{\partial}{\partial \alpha} w_y^* \left(\frac{L^2 - x^2}{2} \right) \tag{9-15}$$

$$N_\alpha = -R w_z$$

The normal forces N_α, N_x and shear force $N_{x\alpha}$ for the most commonly used curvature shapes—ellipse, circle, and cycloid—will be given for dead and live loads.

Cross section. Ellipse: (Fig. 9-15)

$$R = \frac{a^2 b^2}{(b^2 \sin^2 \alpha + a^2 \cos^2 \alpha)^{3/2}}$$

Dead load:

$$w_x = 0, \ w_y = w_D \sin \alpha, \ w_z = w_D \cos \alpha$$

$$N_\alpha = -w_D a^2 b^2 \frac{\cos \alpha}{(b^2 \sin^2 \alpha + a^2 \cos^2 \alpha)^{3/2}} \tag{9-16}$$

$$N_{x\alpha} = -w_D x \frac{2b^2 + (b^2 - a^2)\cos^2\alpha}{b^2 \sin^2\alpha + a^2 \cos^2\alpha} \sin\alpha$$

$$N_x = -\frac{w_D l^2}{2}\left(1 - \frac{x^2}{l^2}\right) \cdot$$

$$\frac{a^2 b^2 (b^2 \sin^2\alpha + a^2 \cos^2\alpha)^{1/2}}{3a^2 b^2 + 3b^2(a^2 - b^2)\sin^2\alpha - (b^2 \sin^2\alpha + a^2 \cos^2\alpha)^2}$$

For dead load, normal forces and shear distribution diagrams are illustrated in Fig. 9-15.

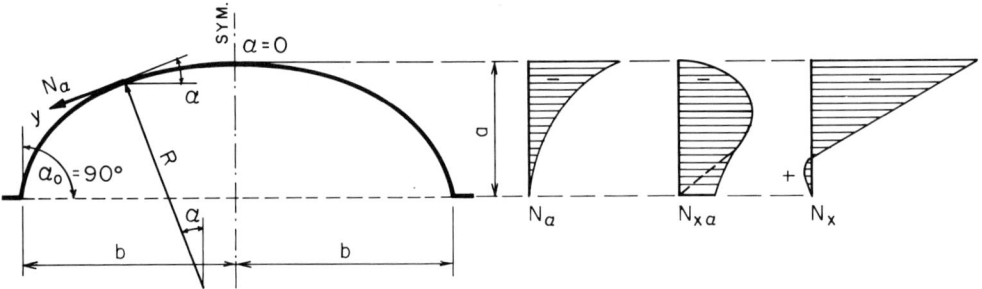

Figure 9-15

Live load:

$$w_x = 0, \quad w_y = w_L \sin\alpha \cos\alpha, \quad w_z = w_L \cos^2\alpha$$

$$N_\alpha = -w_L a^2 b^2 \frac{\cos^2\alpha}{(b^2 \sin^2\alpha + a^2 \cos^2\alpha)^{3/2}}$$

$$N_{x\alpha} = w_L 3x \frac{a^2 - 2(b^2 \sin^2\alpha + a^2 \cos^2\alpha)}{(b^2 \sin^2\alpha + a^2 \cos^2\alpha)} \sin\alpha \cos\alpha \quad (9\text{-}17)$$

$$N_x = -\frac{w_L l^2}{2} 3\left(1 - \frac{x^2}{l^2}\right) \cdot$$

$$\frac{a^2(b^2 \sin^2\alpha - a^2 \cos^2\alpha) + 2(b^2 \sin^2\alpha + b^2 \cos^2\alpha)^2 (\cos^2\alpha - \sin^2\alpha)}{a^2 b^2 (b^2 \sin^2\alpha + a^2 \cos^2\alpha)^{1/2}}$$

Cross section. Circle (Fig. 9-16):
Dead load:

$$\Delta R = R(1 - \sin\alpha_0), \quad z = R(\cos\alpha - \cos\alpha_0)$$

$$w_x = 0, \quad w_y = w_D \sin\alpha, \quad w_z = w_D \cos\alpha$$

$$N_\alpha = -w_D R \cos\alpha$$

Curvilinear Cylindrical Shells (Barrels)

$$N_{x\alpha} = -w_D\, 2x \sin \alpha$$

$$N_x = -\frac{w_D l^2}{R}\left(1 - \frac{x^2}{l^2}\right)\cos \alpha$$

(9-18)

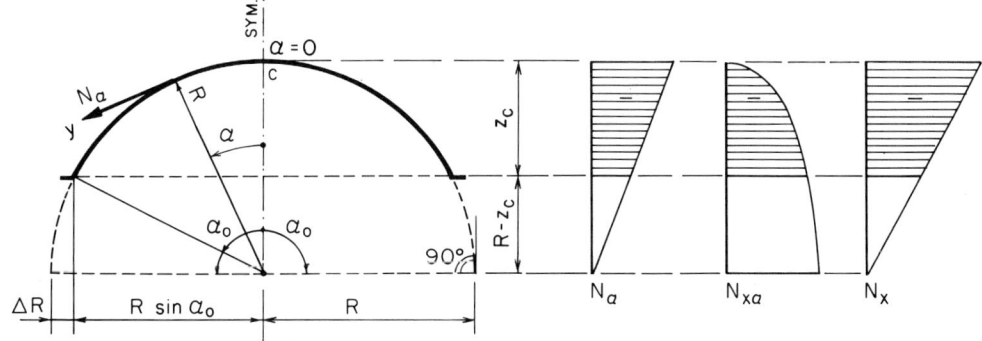

Figure 9-16

Live load:

$$w_\alpha = 0, \quad w_y = w_L \sin \alpha \cos \alpha, \quad w_z = w_L \cos^2 \alpha$$

$$N_\alpha = -w_L R \cos^2 \alpha$$

(9-19)

$$N_{x\alpha} = w_L\, 3x \sin \alpha \cdot \cos \alpha$$

$$N_x = -\frac{w_L l^2}{2}\frac{3}{R}\left(1 - \frac{x^2}{l^2}\right)(1 - 2\sin^2 \alpha)$$

Cross section. Cycloid (Fig. 9-17):

$$z_c = 2r = \frac{a}{2} \qquad a = 2z_c = 4r$$

$$R = a \cos \alpha \qquad \alpha = 90°: \quad R = 0$$

Dead load:

$$w_x = 0, \quad w_y = w_D \sin \alpha, \quad w_z = w_D \cos \alpha$$

$$N_\alpha = -w_D a \cos^2 \alpha$$

$$N_{x\alpha} = -w_D\, 3x \sin \alpha$$

(9-20)

$$N_x = -\frac{w_0 l^2}{2}\frac{3}{a}\left(1 - \frac{x^2}{l^2}\right)$$

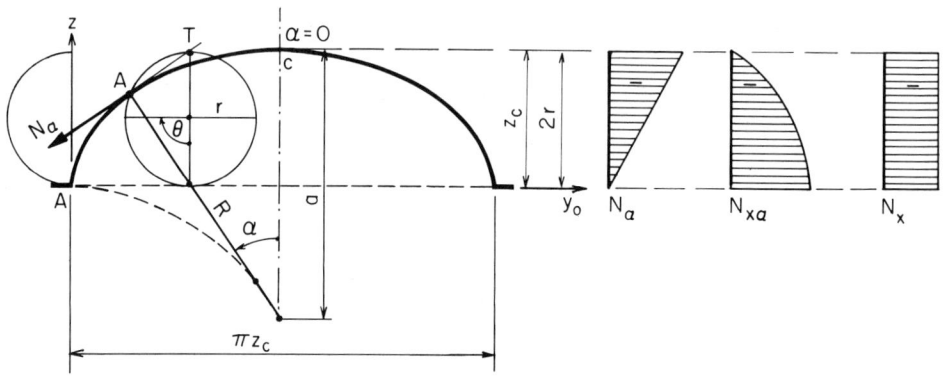

Figure 9-17

Live load:

$$w_x = 0, \quad w_y = w_L \sin \alpha \cos \alpha, \quad w_z = w_L \cos^2 \alpha$$

$$N_\alpha = -w_L a \cos^3 \alpha$$

$$N_{x\alpha} = -w_L 4x \sin \alpha \cos \alpha \qquad (9\text{-}21)$$

$$N_x = -w_L l^2 \frac{2}{a} \left(1 - \frac{x^2}{l^2}\right) \frac{1 - 2\cos^2 \alpha}{\cos \alpha}$$

As can be seen from the foregoing analysis and plotted diagrams, the unbalanced N_α-, $N_{x\alpha}$- and N_x-forces at the shell boundary $\alpha = \alpha_0$ must be carried and balanced by edge members. Also, the equilibrium, in accordance with membrane theory, is possible only in connection with an edge member.

Edge Member

The tensile force T in the edge member at any point x is obtained from the condition

$$T + \sum_{\alpha=0}^{\alpha_0} N_x = 0 \qquad (9\text{-}22)$$

$$T = -\sum_{\alpha=0}^{\alpha_0} N_x$$

The main difficulty with membrane theory lies in balancing the differences in strain and deflection at the junction of the shell and edge beams. Basically, this can be done to a certain degree by the use of prestressing. The prestressing tendons have to balance the T-value and have to be arranged in such a way that $\Sigma \varepsilon$, $\Sigma \delta$ and lateral displacement become approximately zero.

Curvilinear Cylindrical Shells (Barrels)

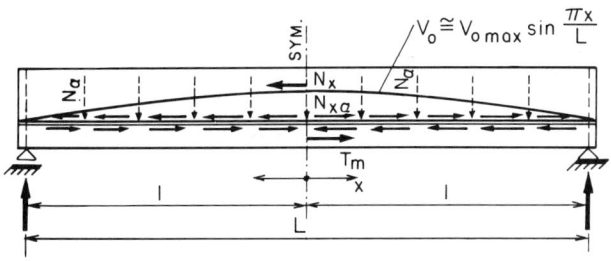
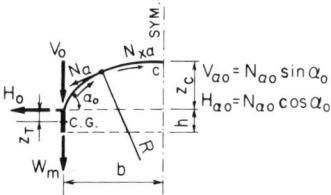

Figure 9-18

The interaction of stresses at the boundary between the shell and beam is shown in Fig. 9-18.

The edge beam acted upon by the unbalanced boundary stresses from the shell (Figs. 9-15, 9-16, and 9-17) causes a sinosoidal moment $\overline{M}_x$ and tensile force T_m in the marginal beam:

$$\overline{M}_{x,\max} = T z_T + \frac{wL^2}{8} \qquad \overline{M}_x = M_{x,\max} \sin \frac{\pi x}{L} \qquad (9\text{-}23)$$

where z_T is the distance from the boundary of shell to the center of gravity of the marginal beam and w is the sum of dead load of the beam and unbalanced N-force.

Beam Theory

As proven by Lundgren, relatively long cylindrical shells and folded plates can be analyzed as beams. The beam method gives qualitatively acceptable results for long shells (lowest limit $L/B > 1.5$). For short cylindrical shells, the distribution of the superimposed σ_x-stresses (shell and edge beam) is not any more approximately linear than long shells.

For example, in the following a long, prestressed shell with cycloidal cross section (Fig. 9-19) will be analyzed by the beam method.

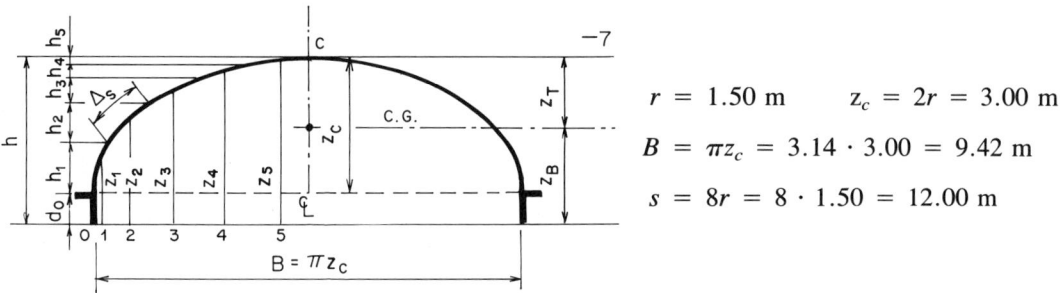

$r = 1.50 \text{ m} \qquad z_c = 2r = 3.00 \text{ m}$

$B = \pi z_c = 3.14 \cdot 3.00 = 9.42 \text{ m}$

$s = 8r = 8 \cdot 1.50 = 12.00 \text{ m}$

Figure 9-19

Geometric data (see "Cycloid," page 191):

$$\Delta s = \frac{s}{n} = \frac{12.00}{10} = 1.20 \text{ m} \qquad t = 0.12 \text{ m}$$

$$z_n = 2r(1 - \cos\vartheta) + d_0$$

$$\Delta A = \Delta s \, t = 1.20 \cdot 0.12 = 0.144 \text{ m}^2$$

Edge Member:

$d_0 = 0.70$ m, $b_0 = 0.20$ m, $\Delta A_m = 0.14$ m^2

$a = 0.40$ m, $d_s = 0.15$ m, $\Delta A_s = 0.06$ m^2

$z_1 = 1.50$ m, $z_2 = 2.35$ m, $z_3 = 2.95$ m, $z_4 = 3.45$ m, $z_5 = 3.70$ m

$h_1 = 1.10$ m, $h_2 = 0.85$ m, $h_3 = 0.56$ m, $h_4 = 0.29$ m, $h_5 = 0.20$ m

$\Sigma h = 3.70$ m

Sectional coefficients (half section):

$$A = 5\,\Delta A + \Delta A_m + \Delta A_s = 5 \cdot 0.144 + 0.14 + 0.06 = 0.92 \text{ m}^2$$

$$Q = \Delta A \Sigma z_n + \Delta A_m z_m + \Delta A_s z_s = 0.144(1.50 + 2.35 + 2.95 + 3.45 + 3.70)$$

$$+ 0.14 \cdot 0.35 + 0.06 \cdot 0.625 = 2.096 \text{ m}^3$$

$$z_B = \frac{Q}{A} = \frac{2.096}{0.92} = 2.28 \text{ m}, \quad z_T = 3.70 - 2.28 = 1.42 \text{ m}$$

$$I_B = \Sigma\left(\frac{\Delta A h^2}{12} + \Delta A z_n^2\right) + \left(\frac{\Delta A_m d_0^2}{12} + \frac{\Delta A_s d_s^2}{12}\right)$$

$$= \frac{0.144}{12}(1.10^2 + 0.85^2 + 0.56^2 + 0.29^2 + 0.20^2)$$

$$+ \frac{0.14 \cdot 0.70^2 + 0.06 \cdot 0.15^2}{12} = 0.034 \text{ m}^4$$

$$+ 0.144\,(1.5^2 + 2.35^2 + 2.95^2 + 3.45^2 + 3.70^2)$$

$$+ 0.14 \cdot 0.35^2 + 0.06 \cdot 0.625^2 = 6.0983 \text{ m}^4$$

$$I_B = 0.0340 + 6.0983 = 6.1323 \text{ m}^4$$

$$I_0 = I_B - A z_B^2 = 6.1323 - 0.92 \cdot 2.28^2 = 1.3498 \text{ m}^4$$

$$i^2 = \frac{I_0}{A} = \frac{1.3498}{0.92} \cong 1.47 \text{ m}^2$$

Curvilinear Cylindrical Shells (Barrels)

As can be seen, the influence of moment of inertia $\Sigma\frac{\Delta A h^2}{12}$ about the center of gravity of the sections (Δs) is less than one percent and can be neglected.

Loading:

Dead load: $\quad W_D = \Sigma A \gamma = 2 \cdot 0.92 \cdot 2.400 = 4.42$ t/m

Live load: $\quad W_L = (B + 2 \cdot 0.40) w_L = 10.22 \cdot 0.150 = 1.53$ t/m

$\quad\quad\quad\quad\quad \Sigma W = \sim 6.0$ t/m

Moment and stresses:

$$M_m = \frac{1}{8}\Sigma w L^2 = \frac{1}{8} \cdot 6.0 \cdot 35^2 \cong 920 \text{ tm}$$

$$\sigma_T = \frac{M}{\Sigma I_0} z_T = \frac{920}{2 \cdot 1.35} 1.42 = -484 \text{ t/m}^2 \cong -48 \text{ kg/cm}^2$$

$$\sigma_B = \frac{M}{\Sigma I_0} z_B = \frac{920}{2 \cdot 1.35} 2.28 = 777 \text{ t/m}^2 \cong 78 \text{ kg/cm}^2$$

Prestressing:

$$e_0 = z_B - d' = 2.28 - 0.48 = 1.80 \text{ m}$$

$$\psi_T = \left(1 - \frac{e_0 z_T}{i^2}\right) = \left(1 - \frac{1.80 \cdot 1.42}{1.47}\right) = -0.74$$

$$\psi_B = \left(1 + \frac{e_0 z_B}{i^2}\right) = \left(1 + \frac{1.80 \cdot 2.28}{1.47}\right) = 3.79$$

$$H_{prn} = \frac{\sigma_B \Sigma A}{\psi_B} = \frac{777 \cdot 1.84}{3.79} \cong 380 \text{ t}$$

$$H_{pr0} = c\, H_{prn} = 1.15 \cdot 380 = 437 \text{ t}$$

$$\sigma_{0pr} = -\frac{H_{pr0}}{\Sigma A} = -\frac{437}{1.84} = -238 \text{ t/m}^2 \sim -24 \text{ kg/cm}^2$$

$$\sigma_{Tpr} = \sigma_{0pr}\, \psi_T = 24 \cdot 0.74 = \sim 18 \text{ kg/cm}^2$$

$$\sigma_{Bpr} = \sigma_{0pr}\, \psi_B = -24 \cdot 3.79 = -91 \text{ kg/cm}^2$$

$$\Sigma\sigma_T = \sigma_T + \sigma_{Tpr} = -48 + 18 = -30 \text{ kg/cm}^2$$

$$\Sigma\sigma_B = \sigma_B - \sigma_{Bpr} = 78 - 91 = -13 \text{ kg/cm}^2$$

The stresses are illustrated in Fig. 9-20.

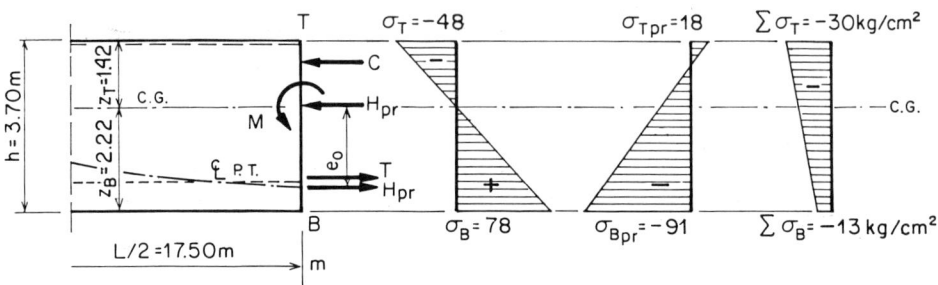

Figure 9-20

Folded plates. By the use of beam method, all shapes of folded plates, subjected to any conceivable loading, can be easily analyzed.

For example (Fig. 9-21):
Geometric data:

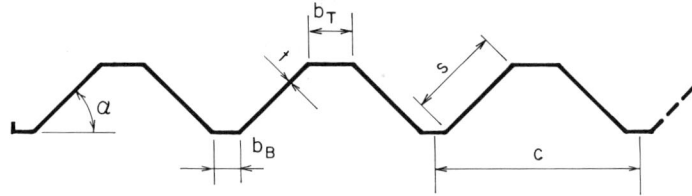

Figure 9-21

$$b_T = 0.54 \text{ m}, \ b_B = 0.40 \text{ m}$$

$$t = 0.12 \text{ m}$$

$$L = 25 \text{ m}$$

$$h = \frac{L}{18} = \frac{25}{18} = 1.40 \text{ m}$$

$$\alpha = 45°$$

$$s = \frac{h}{\sin \alpha} = \frac{1.40}{0.707} = \sim 2.00 \text{ m}$$

$$c = 2s \cos \alpha + b_T + b_B$$

$$= 2 \cdot 2.00 \cdot 0.707 + 0.54 + 0.40 = 3.77 \text{ m}$$

Structural coefficients:

$$A = 2 \cdot 2.00 \cdot 0.12 + 0.54 \cdot 0.12 + 0.40 \cdot 0.12$$

$$= 0.48 + 0.065 + 0.048 = \sim 0.60 \text{ m}^2$$

Curvilinear Cylindrical Shells (Barrels)

$$Q_B = \Sigma \Delta A \cdot z_0 = 0.48 \cdot \frac{1.40}{2} + 0.065 \cdot 1.34 + 0.048 \cdot 0.06 = 0.426 \text{ m}^3$$

$$z_B = \frac{Q_B}{A} = \frac{0.426}{0.60} = 0.71 \text{ m}, \ z_T = 0.64 \text{ m}$$

$$I_0 = \Sigma \left(\frac{\Delta A h^2}{12} + \Delta A z^2 \right)$$

$$= \frac{0.48 \cdot 1.40^2}{12} + \frac{0.065 \cdot 0.12^2}{12} + \frac{0.048 \cdot 0.12^2}{12} = 0.0786 \text{ m}^4$$

$$+ 0.48 \cdot 0.01^2 + 0.065 \cdot 0.63^2 + 0.048 \cdot 0.65^2 = 0.0460 \text{ m}^4$$

$$I_0 = 0.0786 + 0.0460 = 0.1246 \text{ m}^4$$

$$i^2 = \frac{I_0}{A} = \frac{0.1246}{0.60} \cong 0.21 \text{ m}^2$$

Loading:

$$\text{D.L.} = A\gamma = 0.60 \cdot 2.400 = 1.44 \text{ t/m}$$
$$\text{L.L.} = c \, w_L = 3.77 \cdot 0.175 = 0.66 \text{ t/m}$$

$$w = 2.10 \text{ t/m}$$

Moments and reactions:

$$M_m = \frac{1}{8} w L^2 = \frac{1}{8} \cdot 2.10 \cdot 25^2 = 164 \text{ tm}$$

$$R = \frac{wL}{2} = \frac{2.10 \cdot 25}{2} = 26.25 \text{ t}$$

$$\sigma_T = -\frac{M_m}{I_0} z_T = -\frac{164}{0.1246} 0.69 = -908 \text{ t/m}^2 \sim 91 \text{ kg/cm}^2$$

$$\sigma_B = \frac{M_m}{I_0} z_B = \frac{164}{0.1246} 0.71 = 935 \text{ t/m}^2 \sim 94 \text{ kg/cm}^2$$

Prestressing:

$$e_0 = z_B - d' = 0.71 - 0.06 = 0.65 \text{ m}$$

$$\psi_T = \left(1 - \frac{e_0 z_T}{i^2}\right) = \left(1 - \frac{0.65 \cdot 0.69}{0.21}\right) = -1.14$$

$$\psi_B = \left(1 + \frac{e_0 z_B}{i^2}\right) = \left(1 + \frac{0.65 \cdot 0.71}{0.21}\right) = 3.20$$

$$H_{prn} = \frac{\sigma_B \Sigma A}{\psi_B} = \frac{935 \cdot 0.60}{3.20} = 175 \text{ t}$$

$$H_{pr0} = c_L H_{prn} = 1.15 \cdot 175 \cong 200 \text{ t}$$

$$\sigma_{0pr} = -\frac{H_{pr0}}{\Sigma A} = -\frac{200}{0.60} = -333 \text{ t/m}^2 \sim -33 \text{ kg/cm}^2$$

$$\sigma_{Tpr} = \sigma_{0pr}\psi_T = 33 \cdot 1.14 = 38 \text{ kg/cm}^2$$

$$\sigma_{Bpr} = -\sigma_{0pr}\psi_B = -33 \cdot 3.20 = -106 \text{ kg/cm}^2$$

$$\Sigma\sigma_T = -\sigma_T + \sigma_{Tpr} = -91 + 38 = -53 \text{ kg/cm}^2$$

$$\Sigma\sigma_B = \sigma_B - \sigma_{Bpr} = +94 - 106 = -12 \text{ kg/cm}^2$$

The stresses are illustrated in Fig. 9-22.

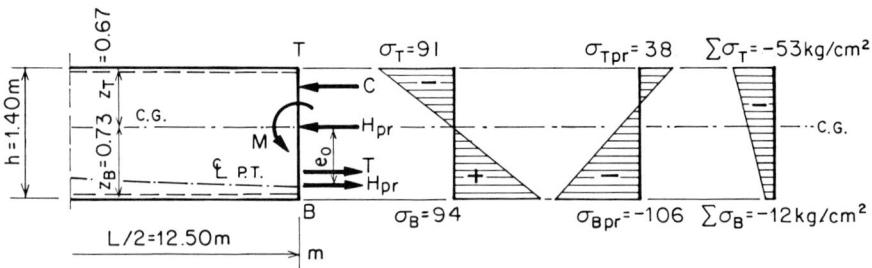

Figure 9-22

Factor of safety:

$$M_{uL} = 1.8 \cdot 164 = 295 \text{ tm}$$

$$T_{uL} = \frac{M_{uL}}{z_0} = \frac{295}{1.15} = 257 \text{ t}$$

$$\Delta T = T_{uL} - 1.05 H_{pr0} = 257 - 1.05 \cdot 200 = 47 \text{ t}$$

$$A_s = \frac{\Delta T}{\sigma_{sy}} = \frac{47}{4.0} = 11.75 \text{ cm}^2$$

POLYGONAL DOMES

Membrane Theory

Polygonal domes are composed of cylindrical barrel sections, which intersections form ridges. Horizontal sections of the dome are polygons. Some of the polygonal dome types are shown in Fig. 9-23.

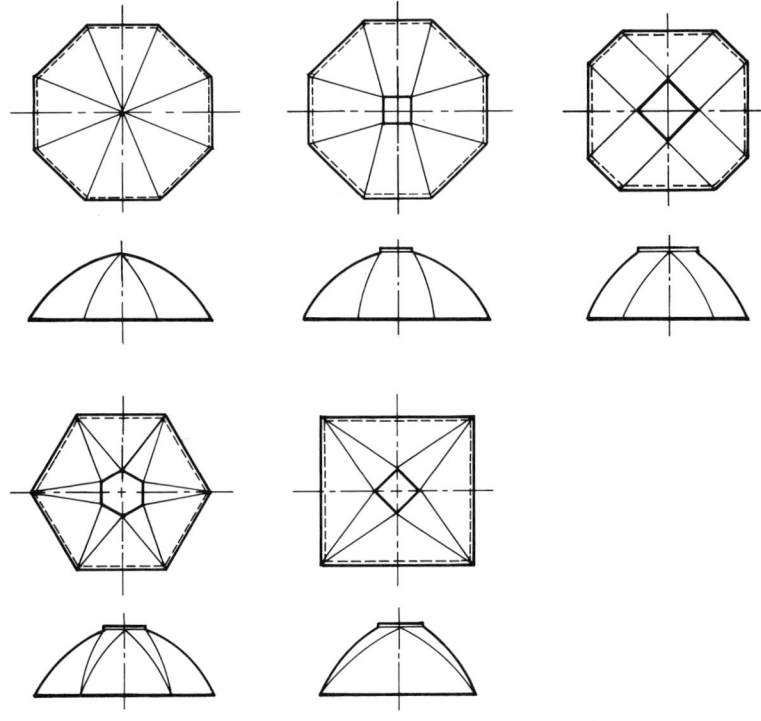

Figure 9-23

The number of sides of a polygon is commonly noted by n and the angle between the ridges by $2\varphi = \dfrac{\pi}{n}$. The transversal section of the barrels can be a circle, cycloid, or ellipse. Generally, a circle is preferred because of easier mathematical treatment of the dome stress condition. The barrel sections between the ridges act primarily as beams. If $\alpha_0 = 90°$, the total load the dome is subjected to is transmitted by membrane forces (N_α, $N_{x\alpha}$ and N_x) over the ridges into the supports (Fig. 9-24).

However, at the boundary ($\alpha_0 < 90°$) the dome must be supported so that the edge displacements and rotations can occur without restrictions. Also, the equilibrium requirements of the forces at the boundary must be met. If $\alpha_0 < 90°$, the boundary forces can be balanced only by the use of the edge member.

The ridge force S_α can be computed from the condition: sum total of the vertical forces at any horizontal section of the dome must be zero—Gulden's rule (Fig. 9-2).

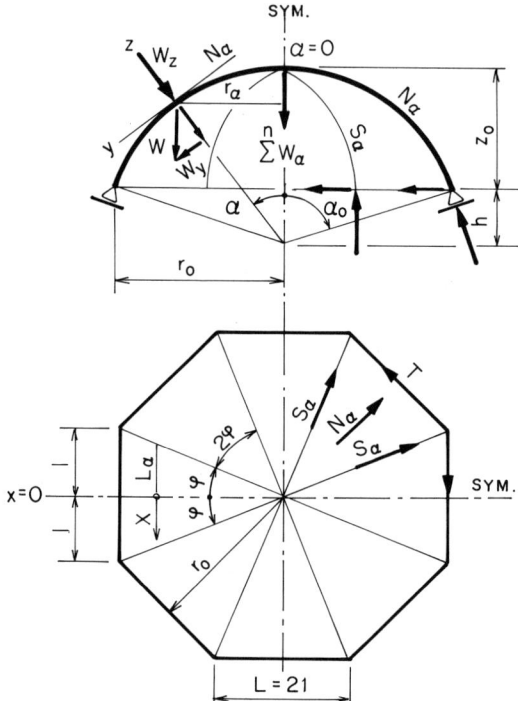

Figure 9-24

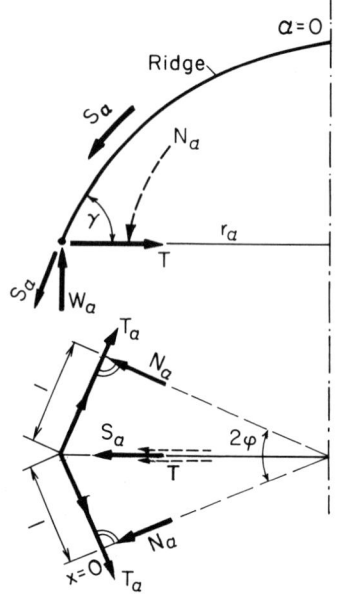

$$\sum_{}^{n} w_\alpha + 2nl_\alpha N_\alpha \sin_\alpha + nS_\alpha \sin \gamma = 0$$

$$l_\alpha = r_\alpha \tan \varphi = R \sin \alpha \tan \varphi$$

$$N_\alpha = w_z R$$

$$\sin \gamma = 1/\sqrt{1 + \cot^2 \alpha/\cos^2 \varphi}$$

(9-24)

Figure 9-25

Polygonal Domes

$$S_\alpha = \left(\frac{\sum_{}^{n} w_\alpha}{n} - 2w_z R r_\alpha \sin \alpha \tan \varphi\right) \frac{1}{\sin \gamma}$$

Dead load: $w_x = 0$, $w_y = w_D \sin \alpha$, $w_z = w_D \cos \alpha$

Live load: $w_x = 0$, $w_y = w_L \sin \alpha \cos \alpha$, $w_z = w_D \cos^2 \alpha$

Meridian is a circle:

$$\sum_{}^{n} w_{\alpha D} = 2n \tan \varphi \, w_D R^2 (1 - \cos \alpha)$$
$$\sum_{}^{n} w_{\alpha L} = 2n \tan \varphi \, \frac{w_L}{2} R^2 \sin^2 \alpha \qquad (9\text{-}25)$$

Meridian is a cycloid:

$$\sum_{}^{n} w_{\alpha D} = 2n \tan \varphi \, 2 w_D z_c^2 \left(\alpha \sin \alpha + \cos \alpha - \frac{1}{3} \cos^3 \alpha - \frac{2}{3}\right)$$
$$\sum_{}^{n} w_{\alpha L} = 2n \tan \varphi \, \frac{w_L}{2} z_c^2 \left(\alpha + \frac{1}{2} \sin 2\alpha\right)^2 \qquad (9\text{-}26)$$

Introducing the $\sum_{}^{n} w_\alpha/n$ from Eqs. 9-25 or 9-26 into Eq. 9-24, we obtain

$$S_\alpha = 2w_\alpha R^2 \tan \varphi [(1 + \sin^2 \alpha) \cos \alpha - 1] \frac{1}{\sin \gamma} \qquad (9\text{-}27)$$

Marginal Member

The equilibrium requires that the sum of horizontal components N_α, $N_{\alpha x}$, N_x and S_α at the corner must be zero. This requires a tension force T_α, which can be computed from the condition

$$T_\alpha + \frac{S_\alpha \cos \gamma}{2 \sin \varphi} + N_\alpha r_\alpha \cos \alpha + \frac{l^2}{2} \frac{\Delta N_{\alpha x}}{\Delta x} \sin \varphi = 0$$

$$T_\alpha = -\left(\frac{S_\alpha \cos \gamma}{2 \sin \varphi} + N_\alpha r_\alpha \cos \alpha + \frac{l^2}{2} \frac{\Delta N_{\alpha x}}{\Delta x} \sin \varphi\right) \qquad (9\text{-}28)$$

$$x = 0: \quad \frac{l^2}{2} \frac{\Delta N_{\alpha x}}{\Delta x} \sin \varphi = 0$$

$$T_\alpha = -\left(\frac{S_\alpha \cos \gamma}{2 \sin \varphi} + N_\alpha r_\alpha \cos \alpha\right)$$

$$N_x = -\frac{\Delta T_\alpha}{\Delta \alpha R}$$

$$= \frac{wR}{\cos^2 \varphi} \left[\frac{1 - \cos \alpha - \cos^2 \alpha}{1 + \cos \alpha} + (1 + 4 \sin^2 \alpha) \cos \alpha \sin^2 \varphi \right] \quad (9\text{-}29)$$

The horizontal forces induce only normal forces into the polygonal dome. However, the N_α-force causes bending in the dome for $\alpha_0 < 90°$. For a symmetrical polygon, the sides with spans $L = 2l$ behave as a fixed beam. Thus, the corner moments

$$M_c = -\frac{L^2}{12} N_\alpha \cos \alpha$$

and in the midspan

$$M_m = \frac{L^2}{24} N_\alpha \cos \alpha \quad (9\text{-}30)$$

The horizontal moments for circular meridian $\alpha < 90°$ the loading is $N_\alpha \sin \alpha$ and for cycloid $N_\alpha = 0$. The vertical moment depends on supporting condition of the polygon.

The forces due to wind (w_w) can be roughly estimated as for rotation symmetrical domes (Eq. 9-5).

Example:

Geometric data:

$$R = 25 \text{ m} \qquad \varphi = \frac{2\pi}{16} = 22.5°$$

$\tan \varphi = 0.414$

$\sin \varphi = 0.383 \qquad \cos \varphi = 0.924$

$\alpha_0 = 72°$

$\sin \alpha_0 = 0.951 \qquad \cos \alpha_0 = 0.309$

$\sin \gamma = 0.899 \qquad \gamma = 64°$

$\cos \gamma = 0.438$

$l = R \sin \alpha_0 \tan \varphi = 9.8 \text{ m}$

$r_0 = R \sin \alpha_0 = 23.7 \text{ m}$

$h = R \cos \alpha_0 = 7.7 \text{ m}$

$z_c = R - h = R(1 - \cos \alpha_0) = 17.3 \text{ m}$

Loading:

$w_D = 0.240 \text{ t/m}^2, \; w_z = w_D \cos \alpha, \; w_y = w_D \sin \alpha$

Polygonal Domes

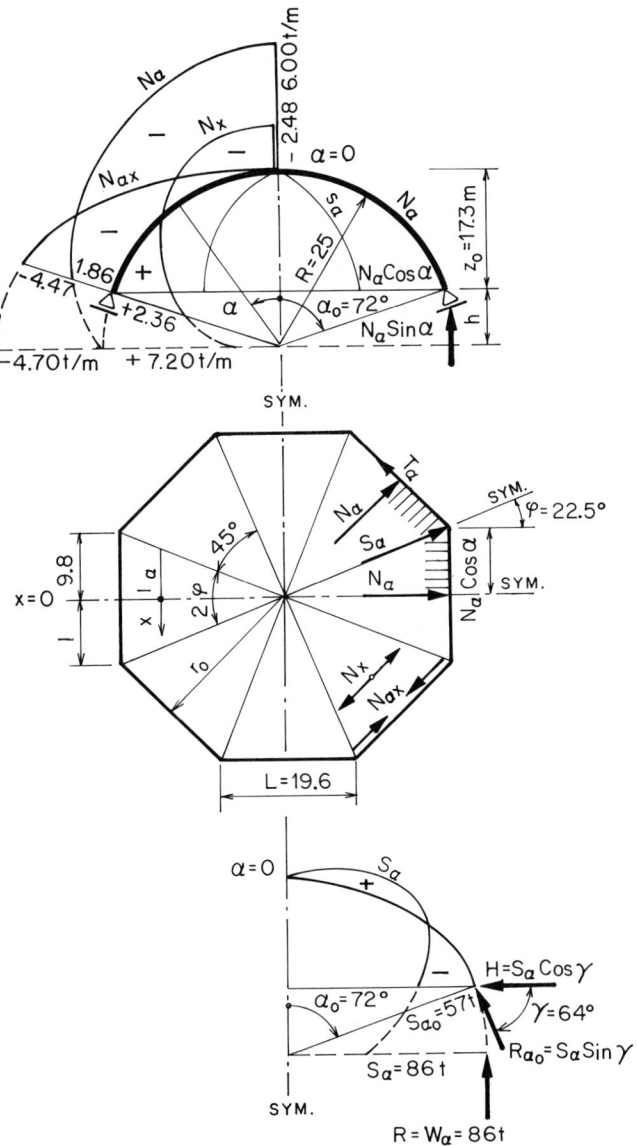

Figure 9-26

$$\sum_{}^{n} w_{\alpha_0} = 2n \tan \varphi \; w_D \; R^2 (1 - \cos \varphi_0) = 686 \text{ t}$$

$$w_{\alpha_0} = \frac{\sum_{}^{n} w\alpha_0}{n} = \frac{686}{8} = 8.6 \text{ t}$$

$\alpha = 0$: $\cos \alpha = 1, \sin \alpha = 0, \sin \gamma = 0$

$$N_\alpha = -w_D R \cos \alpha = -6.00 \text{ t/m}$$

$$S_\alpha = 2w_D R^2 \tan \varphi [(1 + \sin^2 \alpha) \cos \alpha - 1] \frac{1}{\sin \gamma} = 0$$

$$N_{\alpha x} = -2w_D\, x \sin \alpha = 0$$

$$N_x = \frac{w_D R}{\cos^2 \varphi} \left[\frac{1 - \cos \alpha - \cos^2 \alpha}{1 + \cos \alpha} + (1 - 4 \sin^2 \alpha) \cos \alpha \sin^2 \varphi \right]$$

$$\cong -2.48 \text{ t/m}$$

$\alpha_0 = 72°$:

$$N_{\alpha_0} = 0.240 \cdot 25 \cdot 0.309 = -1.86 \text{ t/m}$$

$$S_{\alpha_0} = 2 \cdot 0.240 \cdot 25^2 \cdot 0.414\, [(1 + 0.951^2)0.309 - 1] \frac{1}{0.898}$$

$$= -57 \text{ t}$$

$$N_{\alpha x} = -2 \cdot 0.240 \cdot 9.8 \cdot 0.951 = -4.47 \text{ t/m}$$

$x = 0$: $N_{\alpha x} = 0$

$$N_x = \frac{0.240 \cdot 25}{0.924^2} \left[\frac{1 - 0.309 - 0.309^2}{1 + 0.309} \right.$$

$$\left. + (1 - 4 \cdot 0.951^2)0.309 \cdot 0.385^2 \right]$$

$$= 2.36 \text{ t/m}$$

$$T_{\alpha_0} = \frac{S_{\alpha_0} \cos \gamma}{2 \sin \varphi} + r_{\alpha_0} N_{\alpha_0} \cos \alpha_0 \quad \text{tension ring}$$

$$= \frac{57 \cdot 0.438}{2 \cdot 0.383} + 9.8 \cdot 1.86 \cdot 0.309 = 38.23 \text{ t}$$

$$M_c = -\frac{19.6^2}{12} 1.86 \cdot 0.309 = -18.40 \text{ tm}, \quad M_m = 9.20 \text{ tm}$$

$\alpha_0 = 90°$: $\cos \alpha_0 = 0$, $\sin \alpha_0 = 1$, $\sin \gamma = 1$, $\cos \gamma = 0$

$$N_{\alpha_0} = 0$$

$$S_{\alpha_0} = W_{\alpha_0} = -86 \text{ t}$$

Hyperbolic Paraboloids

$$N_x = \frac{w_D R}{\cos^2 \varphi} = \frac{0.240 \cdot 25}{0.924^2} = 7.02 \text{ t/m}$$

$$N_{\alpha x} = -2 w_D l \sin \alpha_0 = -2 \cdot 0.240 \cdot 9.8 \cdot 1.0 = -4.70 \text{ t/m}$$

$$T_{\alpha_0} = \frac{S_{\alpha_0} \cos \gamma}{2 \cdot \sin \varphi} + r_{\alpha_0} N_{\alpha_0} \cos \alpha_0 = 0$$

Since the $T_{\alpha_0} = 0$, theoretically no tension ring at the support is required.

For a polygonal dome of circular meridian subjected to dead load (w_D), the normal forces N_x, N_α, S_α and shear $N_{\alpha x}$ are illustrated in Fig. 9-26.

HYPERBOLIC PARABOLOIDS

Membrane Theory

The carrying action of a hyperbolic paraboloid is accomplished by two sets of identical parabolic curvatures translated to each other and intersecting at right angles. The first set of paraboloids with downward curvature represents the arch action and the second set with upward curvature represents the suspension action (Fig. 9-27).

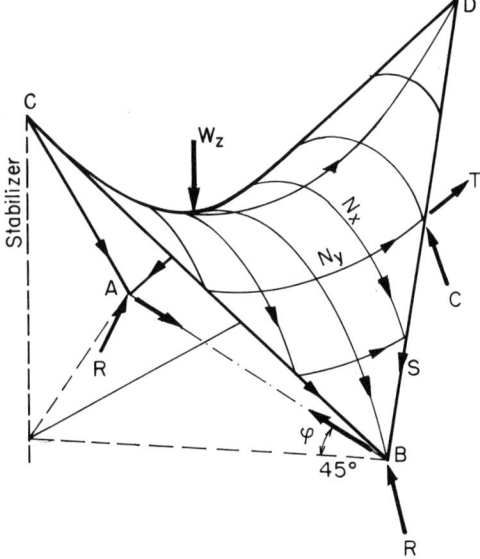

Figure 9-27

The special characteristic of such a translation surface is that a straight line lying wholly in the surface may be drawn through any point on the surface. This property of the hyperbolic paraboloid makes it especially attractive for shell construction.

The general equation for a hyperbolic paraboloid with coordinates as shown in Fig. 9-28 and origin 0 in vertex is

$$z' = \frac{z_a}{a^2} x^2 - \frac{z_b}{b^2} y^2 \tag{9-31}$$

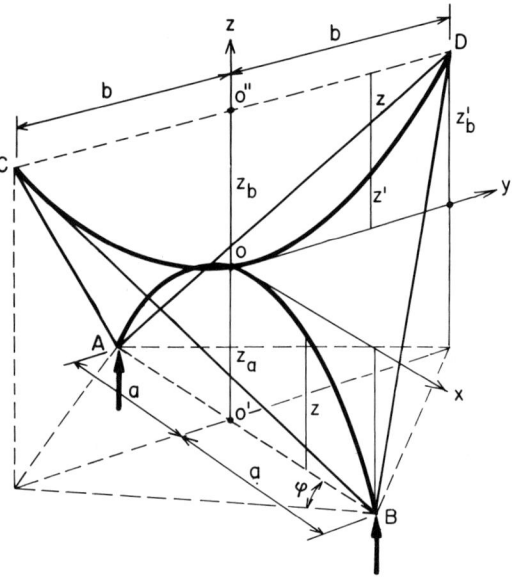

Figure 9-28

In practice, it is more convenient to translate the origins 0 of the individual parabolas to $0'$ and $0''$; then

$$z = z_{a,b} - z'$$

$$z_x = z_a \left(1 - \frac{x^2}{a^2}\right)$$

$$z_y = z_b \left(1 - \frac{y^2}{b^2}\right) \tag{9-32}$$

For a relatively flat parabola $ds \simeq \Delta x$, the slope is $\Delta z/\Delta x$ and the change of slope is $\Delta z^2/\Delta x^2$. Thus, the radius at the vertex turns in the x- and y-direction, respectively

$$R_x = \frac{1}{\Delta z^2/\Delta x^2} = \frac{\Delta x^2}{\Delta z^2} = \frac{a^2}{2z_a}$$

$$R_y = \frac{1}{\Delta z^2/\Delta y^2} = \frac{\Delta y^2}{\Delta z^2} = \frac{b^2}{2z_b} \tag{9-33}$$

Hyperbolic Paraboloids

For identical relatively flat parabolas,

$$R_x = -R_y = R \simeq \frac{ab}{z} \qquad z = z_a + z_b$$

For practical purposes, the R for relatively flat paraboloids can be taken as constant over the entire hyperbolic paraboloidal area.

Substituting $R = \dfrac{ab}{z}$ for R_x and R_y into Eq. 9-33, we obtain the rise of the parabolas respectively

$$z_a = \frac{az}{2b} \qquad z_b = \frac{bz}{2a} \tag{9-34}$$

Membrane Forces

The normal forces N_x and N_y can be computed from the equilibrium condition $\Sigma Z = 0$. Further, considering that for a flat curvature

$$\cos \alpha \simeq 1: \qquad w_z = w_0 \cos \alpha \cong w_D, \; w_Z = w_L \cos^2 \alpha \simeq w_L$$

Thus,

$$N_x + N_y + R w_z = 0$$

$$N_x = -N_y = \frac{w_z}{2} R = w_z \frac{ab}{2z} \tag{9-35}$$

and the twist

$$N_{xy} = 0.$$

The compressive force $N_x = C$ and tensile force $N_y = T$ in the x- and y-directions are the principal forces in the shell and are equivalent and identical to the principal shear S_φ at $\varphi = 45°$ to the x and y axes (principal axes).

Thus,

$$S_\varphi = \pm \frac{1}{2}\sqrt{(N_x - N_y)^2 + 4 N_{xy}^2}$$

$$= \pm \frac{1}{2}\sqrt{(2N)^2 + 0.0} = \pm \frac{w_z}{2} R = \pm w_z \frac{ab}{2z} \tag{9-36}$$

Edge Members

To keep the edge members relatively light, the shell must be supported along straight lines at $\varphi = 45°$ to the principal axes (x, y), where the $N_x = N_y = 0$ and the edge beam is subject to the shear S only. When the shell is supported otherwise, the edge members must balance the resultant of the N_x, N_y-forces at the shell boundary.

The straight edge member of hyperbolic paraboloids is compressed by accumulated shear from C to AB and D to AB, and is also acted upon in its transversal direction by a moment. Thus,

$$P = \Sigma S_\varphi \, \Delta L \, P_{A,Bmax} = w_z \frac{ab}{2z} L \, P_{C,D} = 0, \, M = P \cdot e$$

$$d = \sqrt{a^2 + b^2} \qquad L = \sqrt{d^2 + z^2} \qquad (9\text{-}37)$$

Reactions: $\Delta R_p = P_{max} \sin \alpha$, $\tan \alpha = z/d$, $\Sigma R = \Sigma \Delta R = (4ab/2)w_z$

where $e = b_0/2$ is the eccentricity of the shear in respect to the centerline of the edge member. Also, due to the lack of curvature at the boundary, the shell is not able to carry the dead load of the edge member without developing bending stresses.

To keep the shell free of bending stresses, the edge member deflection in the transversal direction due to M_T and in its plane due to M_D must be avoided. This is done most easily by posttensioning.

Since the total external vertical load $\Sigma A w_z$ is transmitted to the foundation or to supporting elements mainly at A and B, at least one more supporting element is required for stability of the shell to resist wind pressure. This supporting element, or stabilizer, can be most effectively located at point C or D. If this is not possible for architectural or other reasons, then, to avoid torsion and undue stress concentration, two stabilizers are required, arranged symmetrically (if possible) along the edge members $C - AB$ or $D - AB$.

Example (Fig. 9-28):

$$z_a = 3.5 \, m, \, z_b = 4.5 \, m, \, z = z_a + z_b = 8.0 \, m$$

$$a = 12.0 \, m, \, b = 10.0 \, m$$

$$w = w_D + w_L = 0.400 \, t/m^2, \, (w = w_z)$$

$$R_x = -R_y = \frac{a, b}{za} = \frac{12.0 \cdot 10.0}{8.0} = 15.0 \, t = R$$

$$N_x = N_y = \frac{w}{2} R = \frac{0.400}{2} \cdot 15.0 = \mp 3.00 \, t/m$$

$$S_{\varphi = 45°} = N = \pm 3.00 \, t \, m$$

$$d = \sqrt{a^2 + b^2} = \sqrt{12.0^2 + 10.0^2} = 15.6 \, m$$

$$L = \sqrt{d^2 + z^2} = \sqrt{15.6^2 + 8.0^2} = 17.6 \, m \qquad h = z$$

$$P_{max} = S\varphi L = -3.00 \cdot 17.6 = -52.8 \, t$$

$$\tan \alpha = z/d = 8.00/15.6 = 0.513, \, \alpha \cong 27° \, \sin \alpha = 0.454$$

$$\Delta R = P_{max} \sin \alpha = 52.8 \cdot 0.454 \cong 24 \, t$$

Hyperbolic Paraboloids

$$\Sigma R = 4\Delta R = 4 \cdot 24 = 96 \text{ t} = wA = 0.400 \cdot \frac{24 \cdot 20}{2} = 96 \text{ t}$$

$$R_A = R_B = \frac{1}{2}\Sigma R = \frac{1}{2} \cdot 96 = 48 \text{ t}$$

$$H = R_B \tan \alpha = 48 \cdot 0.513 = 24.6 \text{ t}$$

10

Structures Subjected to Dynamic Forces

GENERAL

In areas where dynamic forces occur, the degree of elasticity and ductility of any structure, and especially of high-rise structures, are very important, because the magnitude of dynamic forces is a function of acceleration of a structural system. For rigid structural systems, the acceleration is very high and the ductility is often limited. Therefore, the most damage in earthquake areas occurs in such building ranges. For elastic structural systems having high ductility, the dynamic forces are considerably reduced because of relatively small acceleration. Therefore, the damage due to earthquake and hurricane forces, even on severe occasions, is relatively small.

Presently in structural engineering, the magnitude of elasticity and ductility of a structural system can be determined in wide limits. However, the mathematical treatment to determine the magnitude and duration of the dynamic forces is complex and their limiting values only can be roughly estimated.

The present elastic-plastic highly mathematical computation method, even if able to be applied to complex systems, may give results whose values are highly questionable and misleading.

Therefore, seismic codes have been developed to determine an equivalent static lateral force, as a substitute for the dynamic force, for which the structure should be designed.

General   211

Code Method

The most recent SEAOC* Code does not determine the lateral force at each level, but simply the total seismic force at the base of the structure. The base shear is computed from the equation

$$V = ZKCW \qquad (10\text{-}1)$$

where Z is a zone factor. Here, zone 1-0.25, zone 2-0.5, zone 3-1.0, W is the total weight of the building, and C is a numerical coefficient as a function of the fundamental period (T) of the structure. The C-value can be estimated by

$$C = \frac{0.05}{\sqrt[3]{T}} \qquad (10\text{-}2)$$

where $C = 0.10$ can be assumed for all one- and two-story buildings.

The fundamental period T could be determined in seconds by

$$T = \frac{0.05\,H}{\sqrt{D}} \qquad (10\text{-}3)$$

where H is the total height of the building above base and D is the width of the building. K-value depends mainly on the ductility and energy capacity of the structure. It varies from 0.67 to 1.33 for buildings having structural framing capable of resisting 100 percent of lateral force and buildings with a box system composed of interconnected members.

The lateral force V_z at each level z equals the mass of the level $m_z = \dfrac{W_z}{g}$ times the acceleration α_z. The acceleration of a high-rise building increases greatly over the height. To account for the increased acceleration, the code specifies a triangular base shear distribution over the height of a building. Thus, the lateral force F_z at any level is

$$F_z = V\,\frac{W_z h_z}{\Sigma W_z h_z} \qquad (10\text{-}4)$$

in which W_z is the weight of the level and h_z is the height of level z above the base, and $\Sigma W_z h_z$ is the sum of all $W_z h_z$ from top to the base of the building.

After the structure is designed in accordance with the code, it is advisable to check the adequacy of the structure by the Reserve Energy Method.**

The code method is simple to apply, but there have been several instances of damage to structures designed by this method.

*Structural Engineering Association of California.

**_Design of Multistory Reinforced Concrete Buildings for Earthquake Motions_, Blume Newmark Corning, Portland Cement Association, Chicago, Illinois, 1961.

Energy Method

The energy method is based on the use of elastic acceleration spectral response for any earthquake density F or for any average earthquake spectrum available. Thus, the feed-in kinetic energy from the ground motion can be computed and must be absorbed by the potential energy of the structure. The energy-absorption capacity of a structure depends largely on the flexibility of the system and clearly indicates the range of deformations, safety, and damage to the structure as a function of earthquake or wind impulses.

The energy equation is

$$E_f = E_{s,\max} - E_r \tag{10-5}$$

wherein E_f is feed-in kinetic energy or energy demand, $E_{s,\max}$ is the total energy absorption capacity within the stability range and E_r is the reserve energy of the structure. If $E_r = 0$ for a very severe earthquake intensity, the building would not collapse but may have large unacceptable deformations or even severe damages. For economical reasons, a certain amount of damage is acceptable, thus the E_r is also an indicator of the degree of damage for a certain intensity of earthquake impulse.

The kinetic energy demand per single mass is:

$$E_f = \frac{Wv^2}{2g} = \frac{W}{2g}\frac{a^2g^2T^2}{4\pi^2} \cong \frac{g}{80}a^2T^2W \tag{10-6}$$

$$v = \frac{agT}{2\pi}, \quad a = \frac{2\pi v}{gT}$$

where v is the velocity and a is the acceleration of the mass $\left(\frac{W}{g}\right)$. The v, a-values can be obtained from spectral diagrams, which are available from numerous earthquakes. From these diagrams, simple formulas have been developed to provide basic information for a design.

For 5 percent of critical damping, the spectral velocity is expressed as

$$v_1 = F(T_i)^{1/4} \tag{10-7}$$

and spectral acceleration for the same period (T_i) of 0.30 up to 3.0 seconds

$$a_i = 0.194F(T_i)^{-3/4} \tag{10-8}$$

Estimated values for F and the number of severe earthquake occurrences per 100-year period are: $F = 2.77$, two occurrences; $F = 1.83$, twenty occurrences.

The energy capacity of the structure is determined from the lateral displacements U_n (see "Deflections"). The energy capacity diagram for a single-mass system or story is given in Fig. 10-1.

$$K_D = \frac{V_D}{U_D} \quad \text{(stiffness at } V_D\text{)}$$

General

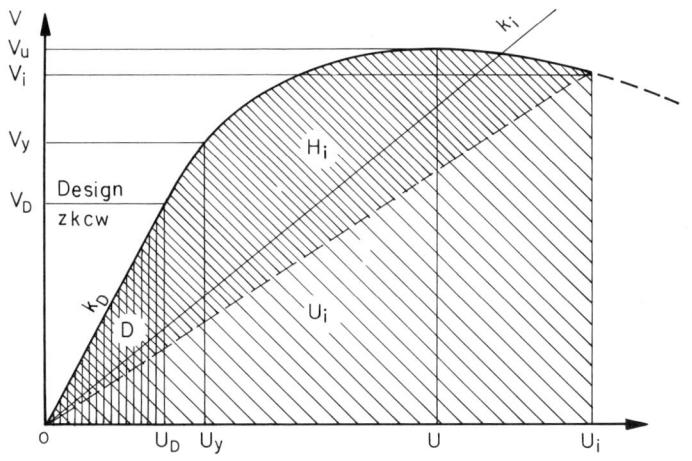

Figure 10-1

$$K_i = \frac{2(U_i - \gamma H)}{U_i^2} \text{ (stiffness at } V_i\text{)}$$

H_i (hump area)

γ (deterioration factor) (10-9)

$$T_i = T\sqrt{K_D/K_i}$$

$$T = 2\pi\sqrt{m/K}$$

$$E_{s,\max} = U_i - \gamma H_i$$

$$E_r = E_{s,\max} - E_f$$

The safety margin (B) of the system considered is

$$B = \sqrt{\frac{E_{s,\max}}{E_f}} = \sqrt{\frac{80(U_i - \gamma H_i)}{ga^2 T_i^2 W}} \qquad (10\text{-}10)$$

Assuming that the equivalent static design coefficient (Eq. 10-2) $C = a_i R_i$, then

$$B = \frac{C}{a_i R_i} \qquad (10\text{-}11)$$

where

$$R_i = \sqrt{\frac{\beta^2}{2\mu - 1}} = \sqrt{\frac{D}{U_i}}$$

is the reserve energy reduction factor, β = ratio of design force (V_D) to yield force

(V_y), and $\mu = \dfrac{U_i}{U_y}$ the ductility factor.

If $R_i a_i$ equals C, the structure or floor under consideration is adequate but has no safety factor at lateral displacement U_i. If $R_i a_i > C$, the structure is inadequate up to the displacement U_i for the specified earthquake intensity F. If $R_i a_i < C$, the structure is safe, provided that the damage (γH_i) and lateral displacement are acceptable. If not, the design must be modified by increasing C or U_i, or both.

Contemporary structures can be designed so that there will be no damage ($\gamma H_i = 0$ and $T_i \to T$) under repeated cycles. Here, the C-value can be determined for any ductility factor μ by $R_i a_i$. The μ-value can be considerably increased by providing adequate elongation length of the reinforcing or prestressing tendons, and by using neoprene interfaces between the column ends and beams (Fig. 10-2).

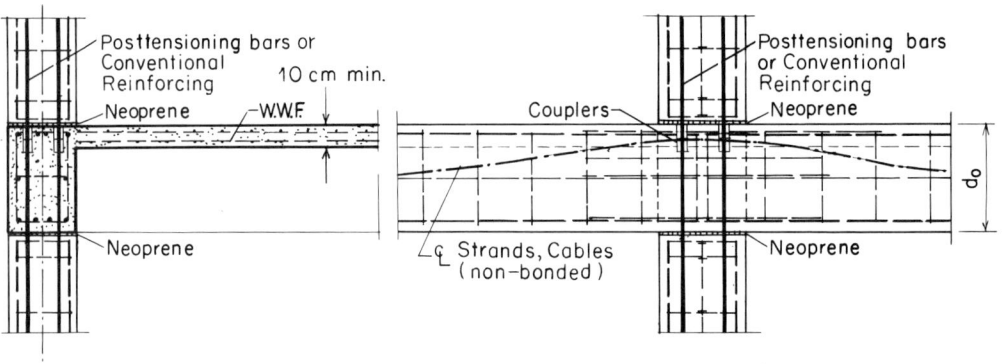

Figure 10-2

Lateral displacement in this case depends on the thickness of the neoprene and provided elongation for steel. Also, the neoprene interfaces serve simultaneously as shock absorbers.

For multistory structures, the energy capacity diagrams for each story have to be computed story by story, or at least for certain stories. However, not all stories will participate in inelastic energy absorption simultaneously or to their full capacity. Therefore, the total energy capacity ($U_i - \gamma H_i$) for each story is suggested to be reduced by a factor $\Omega = 0.3$ to 0.5. Thus,

$$B = \sqrt{\dfrac{80(U_i - \gamma H_i)\Omega}{g a_i^2 T_i^2 W}} \qquad (10\text{-}12)$$

and the period

$$T_i = T \sqrt{\dfrac{(ave\ K_D)\Omega}{ave\ K_i}}$$

WIND LOADING

The steady wind pressure (w_w) depends on the velocity of wind (v), the shape of the structure (ζ), and the structure's surface quality. The steady wind velocity increases with the structure's height above ground level. The wind pressure upon a building is estimated by the formula

$$w_w = \zeta \frac{\rho}{g} \frac{v^2}{2} \tag{10-13}$$

where ρ/g is the density of wind, and as an average it is 0.125 kg sec^2/m^4. In locations where dust is in the air or relative humidity is high, this value could be considerably higher. The shape factor ζ can be determined by model testing in wind tunnels. It is smallest for streamlined cross sections. For cylindrical shapes, it is about 0.50 and for cross sections with large drag its value can be as high as 1.6. Thus

$$w_{w,\min} = 0.063 \, v^2 \text{ kg/m}^2$$

$$w_{w,\max} = 0.200 \, v^2 \text{ kg/m}^2, \, v = \text{m/sec}$$

The steady wind pressure is commonly used for stress analysis (w_w). The gust wind pressure (w_G), which is far higher than the steady wind pressure but has only a short duration, is used to investigate stability and ultimate strength of the structure.

The gust wind pressure (hurricane, tornado) acts upon a structure as a periodic dynamic force, which in certain conditions can cause oscillation and even resonance. The gust wind velocities can be as high as 250 klm per hour. To obtain the static load equivalence, the pressure due to dynamic forces has to be increased by magnification factor

$$v = \frac{1}{1 - \dfrac{\omega^2}{\omega_n^2}} \tag{10-14}$$

where $\omega = 2\pi f$ is the circular frequency of wind and $\omega_n = \sqrt{k/m} = \dfrac{1}{f_n}$ is the natural frequency of a structure. The frequency of gusts (f) has been measured from 0.4 to 0.7 per second and period $T = \dfrac{1}{f} = 2.5$ to 1.4 second.

In case $\omega \to \omega_n$, $v \to \infty$ is the resonance condition, it means a steady buildup of the dynamic forces.

For tornado loading, the gust pressure is accompanied by vacuum, which can be far higher than the gust pressure. Vacuum is also associated with hurricane loading, but to a lesser degree. In case of tornado, the direction of vacuum is almost always upwards, but it can also be horizontal. If vacuum is not considered, relatively light roofs are pulled off and walls lose their stability, which usually results in collapse of the building.

Frequently, wind load would produce more severe loading than an earthquake.

Therefore, whichever loading is greater should be used for design.

One of the most important parts of the design, often overlooked, is the foundation design of a structure in a seismic area. If there is a large differential lateral displacement in foundations of columns or walls, the building most certainly will collapse. Also, piles, if used, should never be driven into bedrock, because the bedrock is elastic and reverse movement to the original location occurs. The overlying strata is inelastic and its movement is only unidirectional. Therefore, piles will be sheared off and lose their carrying capacity.

Also, the stability of a structure against overturning should be carefully checked.

COLUMN STRESSES

In order to determine the stresses in columns (critical) due to dynamic loading, the location x of the neutral plane (N.P.) must be determined from equilibrium ($\Sigma Z = 0$, $\Sigma M = 0$).

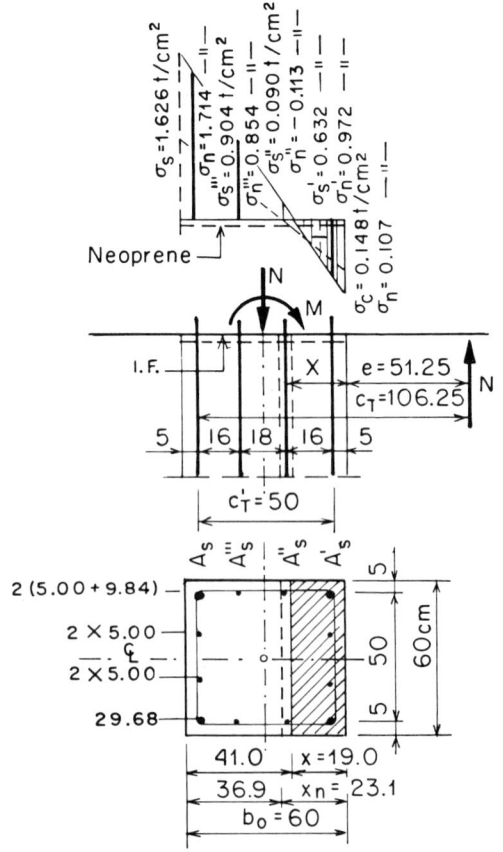

Figure 10-3

Column Stresses

The equilibrium requirement is expressed by Eq. 10-15.

$$x^3 + 3ex^2 + 6(e\overline{A} + \overline{Q})x - 6(e\overline{Q} + \overline{I}) = 0 \qquad (10\text{-}15)$$

where

$$\overline{A} = n(A_s + A'_s)$$

$$\overline{Q} = n(A_s d + A'_s d') \qquad e = \frac{M}{N} - \frac{d}{2}, \text{ (Fig. 10-3)} \qquad (10\text{-}16)$$

$$\overline{I} = n(A_s d^2 + A'_s d'^2)$$

Since the dynamic loads are of relatively short duration, the E_0-value applies. Introducing the $\overline{A}, \overline{Q}, \overline{I}$-values into Eq. 10-15 and solving it for x, the maximum compressive stress σ_c is obtained by Eq. 10-17

$$\sigma_c = -\frac{Nx}{b_0\left(\dfrac{x^2}{2} + x\overline{A} - \overline{Q}\right)} \qquad (10\text{-}17)$$

and the maximum steel stresses by Eq. 10-18

$$\sigma_s = -n\sigma_c \frac{d-x}{x} \qquad (10\text{-}18)$$

$$\sigma'_s = (\sigma_s + n\sigma_c)\frac{c}{d} - \sigma_s$$

The correctness of the computations can be checked by the equilibrium requirement

$$\Sigma Z = 0, \ \Sigma M = 0$$

The final stresses are obtained by superposition of D.L. + 0.25 L.L., $n = 5.8$ and by $n = 11.6$ for dynamic load.

Example

During the period of construction to service, the modulus of elasticity E_0 increased due to further hydration to $E_t = 3.6 \cdot 10^5$ kg/cm^2 by about 5 percent during two months. Thus, for $t_0 \to t$, the n-value becomes

$$n = \frac{E_s}{E_t} = \frac{21.0 \cdot 10^5}{3.60 \cdot 10^5} = \sim 5.8$$

$$\overline{A} = \frac{n(A_s + 2A''_s + A'_s)}{b_0}$$

$$= \frac{5.8(29.68 + 2 \cdot 10 + 29.68)}{60} = 7.67 \text{ cm}^2$$

$$\overline{Q} = \frac{n(A_s d + 2A''_s d'' + A'_s d')}{b_0}$$

$$= \frac{5.8(29.68 \cdot 55 + 2 \cdot 10 \cdot 30 + 29.68 \cdot 5)}{60} = 230 \text{ cm}^3$$

$$\overline{I}_s = \frac{n(A_s d^2 + 2 \cdot A''_s d''^2 + A'_s d'^2)}{b_0}$$

$$= \frac{5.8(29.68 \cdot 55^2 + 2 \cdot 10 \cdot 30^2 + 29.68 \cdot 5^2)}{60} = 10490 \text{ cm}^4$$

$M_{\max} = 36.6$ t m (dynamic)

$N_{\max} = 45.0$ t (dynamic)

$$e = \frac{M_{\max}}{N_{\max}} - \frac{d_0}{2} = \frac{36.6 \cdot 100}{45.0} - \frac{60}{2} = 51.30 \text{ cm}$$

$$x^3 + 3ex^2 + 6(e\overline{A} + \overline{Q})x - 6(e\overline{Q} + \overline{I}) = 0$$

$$x^3 + 154 x^2 + 3741 x - 133734 = 0$$

$$x \cong 19 \text{ cm}$$

$$\sigma_c = \frac{Nx}{b_0\left(\dfrac{x^2}{2} + x\overline{A}_s - \overline{Q}_s\right)} = -\frac{45.0 \cdot 19}{60\left(\dfrac{19^2}{2} + 19 \cdot 7.67 - 230\right)}$$

$$= -0.148 \text{ t/cm}^2$$

$$\sigma_s = n\sigma_c \frac{d-x}{x} = 5.8 \cdot 0.148 \frac{55-19}{19} = 1.626 \text{ t/cm}^2$$

$$\sigma'_s = n\sigma_c \frac{x-d'}{x} = 5.8 \cdot 0.148 \frac{19-5}{19} = -0.632 \text{ t/cm}^2$$

$$\sigma''_s = n\sigma_c \frac{d''-x}{x} = 5.8 \cdot 0.148 \frac{16+5-19}{19} = +0.090 \text{ t/cm}^2$$

$$\sigma'''_s = n\sigma_c \frac{d'''-x}{x} = 5.8 \cdot 0.148 \frac{39-19}{19} = +0.904 \text{ t/cm}^2$$

$$\Sigma Z = N + T + T''' + T'' - (C'_s + C_c)$$

$$= 45.0 + 29.67 \cdot 1.626 + 10.0 \cdot 0.904 + 10.0 \cdot 0.090$$

$$= 45.0 + 48.24 + 9.04 + 0.90 = 103.18 \text{ t}$$

Column Stresses

$$-\left(29.68 \cdot 0.632 + \frac{0.148 \cdot 19 \cdot 60}{2}\right) = -103.12 \text{ t}$$

$$\Sigma M_T = N\, c_T + T\, 0.0 + T'''c_T''' + T''c_T'' - (C_s'c_T' + C_c c_T) \sim 0.0$$

By using elastically controlled joints, the concrete stress is reduced and the steel stresses are increased considerably. The n_N-value for relatively thin (6mm) bearing neoprene interface between columns and beam by the stress-strain diagram (Fig. 3-20) is roughly 11.6. For comparison, the stresses for $n = 11.6$ will be computed in the same section (Fig. 10-3).

The $\bar{A}_s$-, $\bar{Q}_s$- and $\bar{I}_s$-values for $n = 11.6$ are

$$\bar{A}_s = 15.34 \text{ cm}^2,\ \bar{Q}_s = 460 \text{ cm}^3,\ \bar{I}_s = 20980 \text{ cm}^4$$

Introducing these values into Eq. 10-15, we obtain

$$x^3 + 154\, x^2 + 7480\, x - 267468 = 0$$

Solving this equation for x (best by trial) gives roughly

$$x_N = 23.10 \text{ cm},\ \Delta x = x_N - x = 23.10 - 19.0 = 4.10 \text{ cm}$$

$$\sigma_{cN} = -\frac{N \cdot x}{b_0\left(\dfrac{x^2}{2} + x\bar{A}_s - \bar{Q}_s\right)}$$

$$= -\frac{45 \cdot 23.10}{60\left(\dfrac{23.10^2}{2} + 23.10 \cdot 15.34 - 460\right)} = -0.107 \text{ t/cm}^2$$

$$\sigma_{sn} = n\sigma_c \frac{d - x}{x} = 11.6 \cdot 0.107 \frac{55 - 23.10}{23.10} = 1.714 \text{ t/cm}^2$$

$$\sigma_{sN}' = n\sigma_c \frac{x - d'}{x} = -11.6 \cdot 0.107 \frac{23.10 - 5}{23.10} = -0.972 \text{ t/cm}^2$$

$$\sigma_{sN}'' = n\sigma_c \frac{d'' - x}{x} = 11.6 \cdot 0.107 \cdot \frac{21 - 23.10}{23.10} = -0.113 \text{ t/cm}^2$$

$$\sigma_{sN}''' = n \cdot \sigma_c \frac{d''' - x}{x} = 11.6 \cdot 0.107 \frac{39 - 23.10}{23.10} = 0.854 \text{ t/cm}^2$$

For checking:

$$\Sigma Z = N + T + T''' - (C_s'' + C_s' + C_c) \cong 0.0$$

$$\Sigma M = N c_T + T \cdot 0.0 + T'''c_T''' - (T''c_T'' + T'\, c_T' + C_c c_T) = 0.0$$

Differences in stresses for $n = 5.8 \rightarrow 11.6$:

$$\Delta \sigma_c = -0.148 + 0.107 = -0.041 \text{ t/cm}^2$$

$$\Delta\sigma_s = -1.626 + 1.714 = +0.088 \text{ t/cm}^2$$

$$\Delta\sigma_s' = -0.632 + 0.972 = +0.340 \text{ t/cm}^2$$

$$\Delta\sigma_s'' = -0.090 - 0.113 = -0.203 \text{ t/cm}^2$$

$$\Delta\sigma_s''' = -0.904 + 0.854 = -0.050 \text{ t/cm}^2$$

As can be seen from the preceding stress analysis, use of elastically controlled joints causes the x-value to increase from 19.0 cm to 23.1 cm, resulting in reduced stresses in concrete and tensile steel and increased stresses in compressive steel. In addition, the elasticity of a concrete structure is increased considerably and dynamic forces are reduced. There is a large variation in the data concerning the dynamic impulses: periods, frequency, and amplitude, leading to uncertainty.

To overcome these difficulties, a reverse analysis should be used. The design should not start from estimating the dynamic forces, but from the energy absorption capacity of a structure. As can be seen from the energy capacity diagram (Fig. 10-1), because of the increased elongation length for steel the $K_{D,i}$-line remains straight and the distance between V_0 and V_y is increased.

By eliminating the bond between steel and concrete, the elongation length of steel and the thickness-durameter of neoprene interface are controlled by acceptable lateral displacement of the structure.

In accordance with this approach, the structure is analyzed for V_0 (design load). Since the stresses and deformations are proportional to the V-load, the values for higher levels are obtained by multiplication of the stresses by the factor V_n/V_{n-1}. The increase of the E_t-value can be disregarded (added safety) or corrected, if desired. From the obtained results—the stress condition, crack formation, and stability—the maximum lateral force V_{uL} that a structure can withstand safely may be determined.

11

Foundations and Retaining Walls

GENERAL

The bearing capacity of various soils depends on the physical characteristics of soils and also, to a large degree, on the flexibility of footings. Measurements indicate that for every point of the base of a rigid footing, subjected to a centric load, the settlement is the same but the soil pressure is nonuniform. For granular soils, the soil pressure (p) is highest at the center of the footing. In addition, due to lateral displacement of the soil, it decreases parabolically toward the edges of the footing. The soil pressure in plastic soils is higher at the edges because of shear and cohesion. However, in practice uniform soil pressure is used for both soil types for dimensioning of the rigid individual footings. The soil pressure caused by concentrated column load (P) or walls and moments (M) is computed by the formula

$$p = -\frac{P}{A} \pm \frac{M}{I}\frac{b}{2} \qquad (11\text{-}1)$$

where A is the contact area and b is the width of the footing.

For design of foundation in areas where relatively low bearing capacity exists, and also in areas where buildings are subjected to high dynamic loadings (tornados, hurricanes, and earthquakes), utmost care should be taken to avoid relative differential settlements and lateral displacements of footings. This condition can be successfully solved by strip or mat foundations. Further, the column connection with the foundation should not be rigid but elastically controlled. This can be done by using neoprene interfaces between column and foundation and by providing adequate elongation length (l) to dowels by reducing bond between the dowels and the concrete. Also, the neoprene interfaces considerably reduce stress concentration in concrete and soften the dynamic impulses. The elongation length should be such that the stresses in the dowels remain below yield stress (see Fig. 11-1).

Strip and Mat Footings on Semielastic Soil

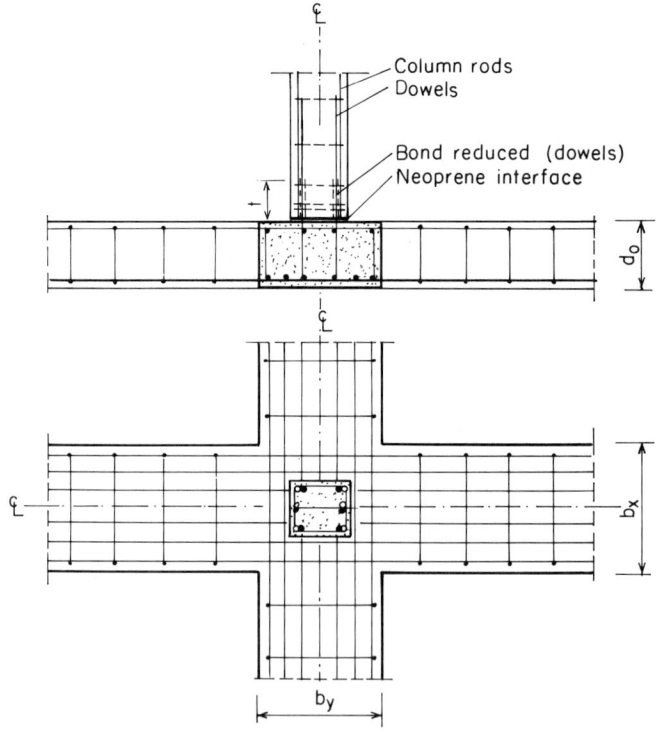

Figure 11-1

The strips and mats have for practical and economical reasons a uniform thickness (d_0) and cannot be considered any more rigid. They should be analyzed as elastic beams or slabs on semielastic soil.

The mats and strips often are computed as flat slabs or continuous beams loaded with uniform soil pressure (p). This method results in considerably higher moments than those computed by the elastic method.

The elastic method is based on the assumption

$$\frac{p}{\delta} = k \text{ (constant)} \tag{11-2}$$

General

where p is the soil pressure, δ the deflection of the beam, and k is the coefficient of subgrade reaction. The relative stiffness L of the beam is

$$L = \sqrt[4]{\frac{4EI}{bk}} \tag{11-3}$$

where E is the modulus of elasticity of concrete, I is the moment of inertia, and b is the width of the beam or strip.

The deflection (δ)* of the beam on semi-elastic subgrade is highest at the point of load application and is damping in accordance with harmonic vibration. For high k-values, the damping of δ is rather rapid. The zero-point distances (x_0) are

$$x_0 = \pi L = \pi \sqrt[4]{\frac{4EI}{bk}} \tag{11-4}$$

The differential equation of the deflection (δ), related moments (M), and shear (V), expressed in exponential function, are for infinitely long beams subjected to concentrated load (P_0) and (M_0) at $\xi = \dfrac{x}{L} = 0$.

P_0 loading

$$\delta_\xi = \frac{P_0}{2Lbk} e^{-\xi} (\cos \xi + \sin \xi)$$

$$M_\xi = \frac{P_0 L}{4} e^{-\xi} (\cos \xi - \sin \xi) \tag{11-5}$$

$$V_\xi = -\frac{P_0}{2} e^{-\xi} \cos \xi$$

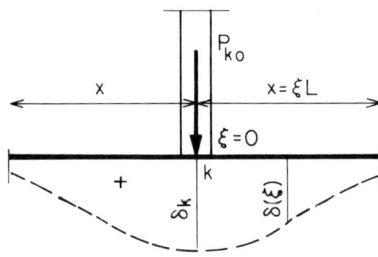

Figure 11-2

*The formulas are taken from the author's book, *Contemporary Concrete Structures* (New York: McGraw-Hill Book Company, 1972).

M_0 loading

$$\delta_\xi = \frac{M_0}{L^2 bk} e^{-\xi} \sin \xi$$

$$M_\xi = \frac{M_0}{2} e^{-\xi} \cos \xi$$

$$V_\xi = -\frac{M_0}{2L} e^{-\xi} (\cos \xi + \sin \xi) \tag{11-6}$$

$$x_0 = \pi L$$

$$p_\xi = k \delta_\xi$$

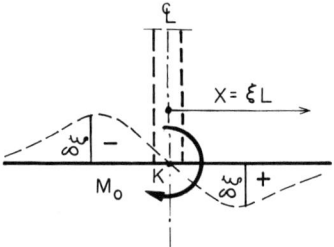

Figure 11-3

For semi-infinitely long beams subjected to (P_0) and (M_0) at their ends $\xi = \dfrac{x}{L} = 0$

$$x_0 = \pi L \mid 2, \ k_0 = \pi L$$

$$p_\xi = k \delta_\xi \tag{11-7}$$

P_0 end loading

$$\delta_\xi = \frac{2P_0}{Lbk} e^{-\xi} \cos \xi$$

$$M_\xi = -P_0 L \, e^{-\xi} \sin \xi \tag{11-8}$$

$$V_\xi = -P_0 \, e^{-\xi} (\cos \xi - \sin \xi)$$

M_0 end loading

$$\delta_\xi = -\frac{2M_0}{L^2 bk} e^{-\xi} (\cos \xi - \sin \xi)$$

$$M_\xi = M_0 \, e^{-\xi} (\cos \xi + \sin \xi) \tag{11-9}$$

$$V_\xi = -\frac{2M_0}{L} e^{-\xi} \sin \xi$$

General

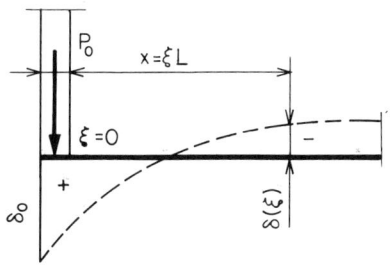
Figure 11-4

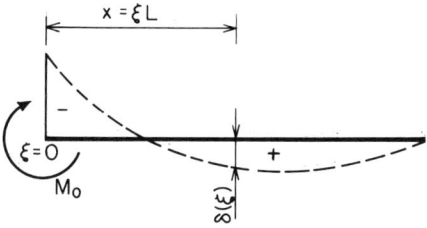
Figure 11-5

In practice, there are no infinitely long beams; therefore, in a finitely long beam we obtain moments M_E and shear V_E at the beam ends. To make them zero in order to satisfy the boundary conditions, the end moments and shear computed by Eqs. 11-5 and 11-6 will be applied as external forces with countersigns in Eqs. 11-8 and 11-9. The final values of $\Sigma\delta$, ΣM, and ΣV will be obtained by one operation only.

Example

$$E = 2{,}50000 = 2.5 \cdot 10^5 \text{ kg/cm}^2$$

$$I = 1{,}040000 = 1.04 \cdot 10^6 \text{ cm}^4$$

$$b = 100 \text{ cm}, \ k = 30 \text{ kg/cm}^3$$

$$P_0 = 100 \ t. \ \xi = 0$$

$$L = \sqrt[4]{\frac{4 \ EI}{b \ k}} = \sqrt[4]{\frac{4 \cdot 2.5 \cdot 10^5 \cdot 1.04 \cdot 10^6}{100 \cdot 30}} = 136 \text{ cm} = 1.36 \text{ m}$$

$$\delta_0 = \frac{P_0}{2Lbk} = \frac{100 \cdot 1000}{2 \cdot 136 \cdot 100 \cdot 30} = 0.123 \text{ cm}, \ (\xi = 0)$$

$$p_0 = k\delta_0 = 30 \cdot 0.123 = 3.70 \text{ kg/cm}^2 = 37.0 \text{ t/m}^2$$

Checking:

$$x_0 = \pi \ L = 3.14 \cdot 1.36 = 4.27 \text{ m}$$

$$P_0 \cong 2/3 \ p_0 \ x_0 = 2/3 \cdot 37.0 \cdot 4.27 \cong 105 \text{ t}$$

Two-way system:

$$p_x = p_y = 1/2 \ p_0 = 18.5 \text{ t/m}^2$$

Moment:

$$M_{x,\max} = M_{y,\max} = \frac{1/2 \ P_0 \ L}{4} = \frac{50 \cdot 1.36}{4} \cong 17 \text{ t m}$$

Reinforcement:

$$d_0 = 50 \text{ cm}, \quad d = 42 \text{ cm} \qquad k_x = \frac{15\sigma_c}{15\sigma_c + \sigma_s} = \frac{15 \cdot 60}{15 \cdot 60 + 1400} = 0.391$$

$$x = k_x d = 0.391 \cdot 42 \cong 16.4 \text{ cm}$$

$$z_0 = d - \frac{x}{3} = 42 - \frac{16.4}{3} \cong 36.5 \text{ cm}$$

$$T = -C = \frac{M_{max}}{z_0} = \frac{17}{0.365} = 46.6 \text{ t}$$

$$A_s = \frac{T}{\sigma_s} = \frac{46.6}{1.400} = 33.3 \text{ cm}^2$$

$$\sigma_c = \frac{2C}{b\,x} = \frac{2 \cdot 46600}{100 \cdot 16.4}\, 1000 = 57 \text{ kg/cm}^2$$

Negative moment:

$$x = 2\pi/2\, L = 3.14 \cdot 1.36 = 4.27 \text{ m}, \quad \xi = \frac{x}{L} = 3.16$$

$$M = \frac{1/2 P_0 L}{4} e^{-\xi} (\cos \xi - \sin \xi)$$

$$\cong \frac{50 \cdot 1.36}{4}\, 0.0434(-0.999 - 0.030) \cong -0.78 \text{ t m}$$

The k-values and allowable soil pressure (p) are presented in Table 4. As can be seen, these values vary to a considerable degree. To be safe, the limiting value approach is recommended. Between the limiting values, minimum and maximum, the foundation will be safe.

In earthquake areas where pile foundation is required, the piles should not be in contact with the bedrock. The rock is elastic and punches back after the earthquake impulse disappears, but the inelastic overlying strata do not. The different relative displacement of rock and overlying strata can damage the piles. The safest are the pedestal piles with heavily reinforced shafts.

RETAINING WALLS

The weight of soil (γ) varies considerably (Table 4), mostly due to the water content in soil. The pressure on a retaining wall consists therefore of earth pressure (p_γ), surcharge (p_s), water pressure (p_w), and reduced earth pressure (p_u) due to uplift. The uplift depends on the volume of soil reduced by voids (ε) in soil $V = (1 - \varepsilon)$. The ε-value of

Retaining Walls

soil depends on density. For sandy soils, it may vary from 30 to 40 percent; for gravel-sandy soil, 20 to 30 percent; and for clay-soil, it is about 30 percent.

In accordance with Coulomb and Poncelet, the active horizontal earth pressure ($E\gamma$) is

$$E_\gamma = \frac{1}{2} \gamma h^2 c \qquad (11\text{-}10)$$

$$c = \tan^2\left(45° - \frac{\varphi}{2}\right)$$

where φ is the natural slope of the soil and h the height of wall.

The pressure (E_s) due to surcharge is

$$E_s = shc \qquad (11\text{-}11)$$

and the water pressure reduced by uplift (p_u)

$$E_w = \frac{h_w^2}{2} - E\gamma \frac{h_w^2}{h^2} \frac{(1 - \varepsilon)}{\gamma} \qquad (11\text{-}12)$$

where h_w is the height of water level.

Free Standing Walls

Example

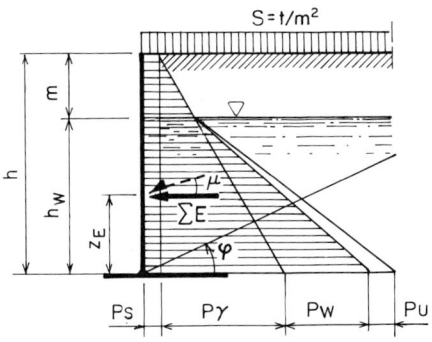

Figure 11-6

$h = 5.00$ m $\qquad \varphi = 20° - 26°$

$h_w = 3.50$ m $\qquad c = 0.422$

$\gamma = 1.8$ t/m^3 $\qquad \varepsilon = 0.35$

$\gamma_w = 1.0$ t/m^3 $\qquad \mu = 17°$

$$\gamma_c = 2.400 \text{ t/m}^3 \quad s = 0.250 \text{ t/m}^2$$

$$E_\gamma = \frac{1}{2}\gamma h^2 c = \frac{1}{2} 1.8 \cdot 5.00^2 \cdot 0.422 = 9.5 \text{ t/m}$$

$$E_v = E_\gamma \tan\mu = 9.5 \cdot 0.287 = 2.73 \text{ t/m}$$

Surcharge:

$$E_s = sh \tan^2\left(45° - \frac{\varphi}{2}\right) = 0.250 \cdot 5.00 \cdot 0.422 = 0.53 \text{ t/m}$$

Water pressure:

$$E_{w+u} = \frac{h_w^2}{2} - E_\gamma \frac{h_w^2}{h^2}\frac{1-\varepsilon}{\gamma} = \frac{3.50^2}{2} - 9.50 \frac{3.50^2}{5.00^2}\frac{1-0.35}{1.8} = 4.44 \text{ t/m}$$

$$p_\gamma = \frac{2E_\gamma}{h} = \frac{2 \cdot 9.5}{5.00} = 3.80 \text{ t/m}^2$$

$$p_s = \frac{E_s}{h} = \frac{0.53}{5.00} = 0.11 \text{ t/m}^2$$

$$p_w = \frac{2 \cdot E_w}{h_w} = \frac{2 \cdot 4.44}{3.50} = 2.53 \text{ t/m}^2$$

$$p_u = \frac{2E_\gamma}{h_w}\frac{h_w^2}{h^2}\frac{1-\varepsilon}{\gamma} = \frac{2 \cdot 9.50}{3.50}\frac{3.50^2}{5.00^2}\frac{1-0.35}{1.8} = 0.96 \text{ t/m}^2$$

$$\Sigma M_E = E_\gamma \frac{h}{3} + E_s \frac{h}{2} + E_w \frac{h_w}{3} = 9.50 \cdot \frac{5.00}{3} + 0.53 \frac{5.00}{2}$$

$$+ 4.44 \frac{3.50}{3} = 22.36 \text{ t m/m}$$

$$\Sigma E = E_\gamma + E_s + E_w = 9.50 + 0.53 + 4.44 = 14.47 \text{ t/m}$$

$$z_0 = \frac{\Sigma M_E}{\Sigma E} = \frac{22.36}{14.47} = 1.55 \text{ m}$$

Footing:

$$\Sigma W_1 = a_2(h_\gamma + s + \varepsilon\gamma_w h_w) + E_v$$

$$= 1.60(5.00 \cdot 1.8 + 0.250 + 0.35 \cdot 1.0 \cdot 3.50) + 2.73$$

$$= 16.76 + 2.73 = 19.49 \text{ t/m}$$

$$W_2 = \frac{d_{0T} + d_{0B}}{2} h \gamma_c = \frac{0.30 + 0.50}{2} 5.00 \cdot 2.400 = 4.80 \text{ t/m}$$

Retaining Walls

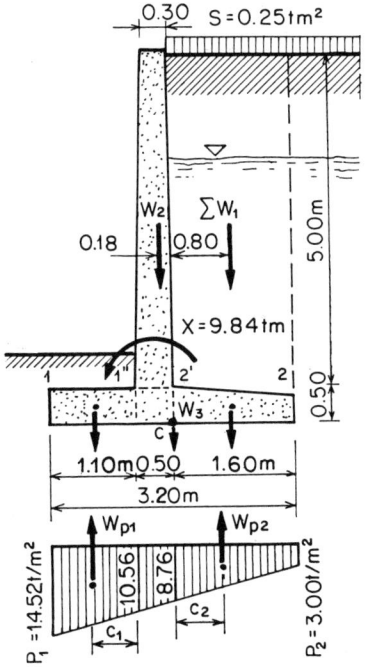

Figure 11-7

$$\Sigma M_c = -(\Sigma M_E + W_2 \cdot 0.183) + \Sigma W_1 \, 0.80 = X$$

$$= -(22.36 + 4.80 \cdot 0.183) + 16.76 \cdot 0.80 = -9.84 \text{ t m/m}$$

$$e_0 = \frac{\Sigma M_c}{\Sigma W} = -\frac{9.84}{19.49 + 4.80} = -0.40 \text{ m}$$

$$A_F = 1.0 \, a = 1.0 \cdot 3.20 = 3.20 \text{ m}^2$$

$$I_F = \frac{A \cdot b^2}{12} = \frac{3.20 \cdot 3.20^2}{12} = 2.73 \text{ m}^4$$

$$W_3 = a \, 1.0 \, \gamma_c = \sim 3.20 \cdot 1.0 \cdot 0.50 \cdot 2.30 \cong 3.70 \text{ t/m}$$

$$p_{1,2} = -\frac{\Sigma W}{A_F} \mp \frac{\Sigma M_c}{I_F} a_c = -\frac{19.49 + 4.80 + 3.70}{3.20}$$

$$\mp \frac{9.84}{2.73} 1.60 = -8.75 \mp 5.77 = -14.52 \text{ t/m}^2, \; -3.00 \text{ t/m}^2$$

$$\text{F.S.} = \frac{16.76 \cdot 2.40 + 4.80 \cdot 1.42 + 2.73 \cdot 1.60 + 3.70 \cdot 1.60}{14.47 \, (1.55 + 0.50)}$$

$$= \frac{57.33}{29.66} = 1.93$$

$$W_{p1} = \frac{14.52 + 10.56}{2} \cdot 1.10 = 13.79 \text{ t/m}$$

$$c_1 = 0.58 \text{ m}$$

$$W_{p2} = \frac{8.76 + 3.00}{2} \cdot 1.60 = 9.41 \text{ t/m}$$

$$c_2 = 0.67 \text{ m}$$

Resistance against sliding:

$$R_s = \Sigma W f = (19.49 + 4.80 + 3.70)0.50 \cong 14 \text{ t/m} < 14.47 \text{ t/m}$$

Moments:

$$M_1'' = W_{P1}\, c_1 = 13.79 \cdot 0.58 = \sim 8.0 \text{ t m/m}, \quad \left(-\frac{3.70}{3.20}\frac{1.25^2}{2} \sim 1.0 \text{ tm/m}\right)$$

$$M_2' = W_{P2}\, c_2 - \left(\Sigma W_1 + \frac{W_3}{2}\right)\frac{a}{2} = -9.41 \cdot 0.67 + \left(16.76 + \frac{3.70}{2}\right)\frac{1.60}{2}$$

$$= -6.30 + 14.88 \simeq 8.6 \text{ t m/m}$$

Reinforcing & stresses: (Eqs. 6-3, 6-4, and 6-5)

Wall:

$$\Sigma M_E = 22.36 \text{ t m} \qquad N = 4.80 \text{ t/m}$$

$$d_0 = 50 \text{ cm}, \quad d = 50 - 8 = 42 \text{ cm}$$

$$k_x = \frac{15 \cdot \sigma_c}{15\sigma_c + \sigma_s} = \frac{15 \cdot 70}{15 \cdot 70 + 1400} = 0.429$$

$$x = k_x d = 0.429 \cdot 42 = \sim 18 \text{ cm}$$

$$z_0 = d - \frac{x}{3} = 42 - \frac{18}{3} = 36 \text{ cm}$$

$$T = -C = \frac{\Sigma M}{z_0} = \frac{22.36}{0.36} = 62 \text{ t}$$

$$M_T = 22.36 + 4.8 \cdot 0.18 = 23.22 \text{ t m/m}$$

$$A_s = \frac{M_T}{z_0 \sigma_s} - \frac{N}{\sigma_s} = \frac{23.22}{0.36 \cdot 1.400} - \frac{4.80}{1.400} = 43.64 \text{ cm}^2$$

Retaining Walls

$$v = \frac{\Sigma E}{b \, z_0} = \frac{14.47 \cdot 1000}{1.00 \cdot 36} = 4 \text{ kg/cm}^2$$

$$\sigma_c = -\frac{2C}{x} - \frac{N}{A} = -\left(\frac{2 \cdot 62}{18} + \frac{4.80}{50}\right)\frac{1000}{100} \cong -70 \text{ kg/cm}^2$$

Footing:

$$M_1'' = 7.0 \text{ t m/m} \qquad d = 42 \text{ cm} \qquad \sigma_c = \sim 30 \text{ kg/cm}^2$$

$$x \cong 10.2 \text{ cm}, \; z_0 = 38.6 \text{ cm}$$

$$T = -C' = \frac{7.0 \cdot 100}{38.6} = 18 \text{ t/m}$$

$$A_s = \frac{18}{1.400} = 12.86 \text{ cm}^2/\text{m} - \text{Bottom}$$

$$\sigma_c = \frac{2 \cdot 18 \cdot 1000}{10.2 \cdot 100} \cong 35 \text{ kg/cm}^2$$

$$M_2' = 8.0 \text{ t m} \qquad d = 42 \text{ cm} \qquad \sigma_c = 35 \text{ kg/cm}^2$$

$$x = 11.5 \text{ cm}, \; z_0 \cong 38 \text{ cm}$$

$$T = -C = \frac{8.0 \cdot 100}{38} = 21 \text{ t/m}$$

$$A_s = \frac{21}{1.400} = 15 \text{ cm}^2/\text{m} - \text{Top}$$

$$\sigma_c = \sim 36 \text{ kg/cm}^2$$

Laterally Supported Walls

$$P = 1.70 \text{ t/m}$$

$$s = 0.500 \text{ t/m}^2 - \text{surcharge}$$

$$E_\gamma = 9.50 \text{ t/m}, \; E_V = 2.73 \text{ t/m}$$

$$E_w = 4.44 \text{ t/m}$$

$$E_s = 0.500 \cdot 5.00 \cdot 0.422 = 1.05 \text{ t/m}$$

$$\Sigma M_E = 9.50 \, \frac{5.00}{3} + 1.05 \cdot \frac{5.00}{2} + 4.44 \, \frac{3.5}{3} = 23.64 \text{ t m/m}$$

$$\Sigma E = 9.50 + 1.05 + 4.44 \cong 15.00 \text{ t/m}$$

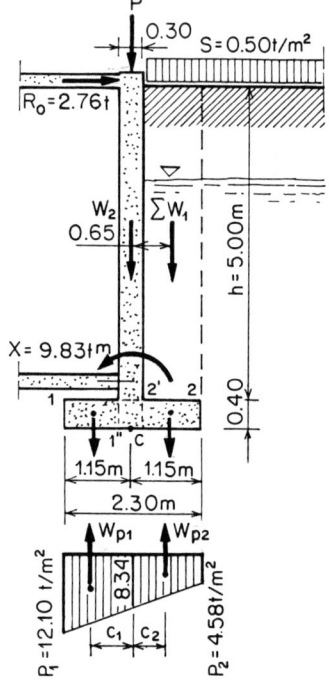

Figure 11-8

$$z_0 = \frac{\Sigma M_E}{\Sigma E} = \frac{23.64}{15.00} = 1.57 \text{ m}$$

$$\Sigma W_1 = (5.00 \cdot 1.8 + 0.500 + 0.35 \cdot 1.0 \cdot 3.50)(1.30 - 0.30)$$

$$= 10.725 \cdot 1.00 = 10.73 \text{ t/m}$$

$$W_2 = 5.00 \cdot 0.30 \cdot 2.400 = 3.60 \text{ t/m}$$

$$E_V = 2.73 \text{ t/m}$$

Loading factors:

$$L'_{ps} = \frac{p_s h^2}{4} = \frac{0.50 \cdot 0.422 \cdot 5.00^2}{4} = 1.32$$

$$L'_{p\gamma} = \frac{8}{60} p\gamma h^2 = \frac{8}{60} \cdot 3.80 \cdot 5.00^2 = 12.67$$

$$L'_{pw} = \frac{p_w h_w^2}{60 \, h^2}(20 \, h^2 - 15 \, h \, h_w + 3 \, h_w^2)$$

$$= \frac{2.53 \cdot 3.50^2}{60 \cdot 5.00^2}(20 \cdot 5.00^2 - 15 \cdot 5.00 \cdot 3.50 + 3 \cdot 3.50^2) = 5.67$$

Retaining Walls

$$\Sigma L' = 1.32 + 12.67 + 5.67 = 19.66 \text{ tm/m}$$

Redundant:

$$\Sigma L' + 2X = 0 \qquad X = \frac{19.66}{2} = -9.83 \text{ tm/m}$$

$$R_{OT} = \frac{p_w h_w^2}{6 h} = \frac{2.53 \cdot 3.50^2}{6 \cdot 5.00} = 1.03 \text{ t/m}$$

$$R_{OB} = \frac{p_w h_w}{6}\left(3 - \frac{h_w}{h}\right) = \frac{2.53 \cdot 3.50}{6}\left(3 - \frac{3.50}{5.00}\right) \cong 3.40 \text{ t/m}$$

$$x_0 = (h - h_w) + h_w\sqrt{\frac{h_w}{3h}}$$

$$= 1.5 + 3.50\sqrt{\frac{3.50}{3 \cdot 5.00}} = 3.19 \text{ m}$$

$$M_{0,w} = R_{OT}\left(x_0 - \frac{h_w}{3}\sqrt{\frac{h_w}{3h}}\right) = 1.03\left(3.19 - \frac{3.50}{3}\sqrt{\frac{3.50}{3 \cdot 5.00}}\right)$$

$$= 2.70 \text{ tm/m}$$

Soil pressure:

$$R_{OT} = \frac{p_\gamma h}{6} = \frac{3.80 \cdot 5.00}{6} = 3.17 \text{ t/m}$$

$$R_{OB} = \frac{p_\gamma h}{3} = \frac{3.80 \cdot 5.00}{3} = 6.33 \text{ t/m}$$

Surcharge:

$$R_{T,B} = \frac{0.50 \cdot 5.00 \cdot 0.422}{2} = 0.53 \text{ t/m}$$

$$\Sigma R_{OT} = 1.03 + 3.17 + 0.53 = 4.73 \text{ t/m}$$

$$\Sigma R_{OB} = 3.40 + 6.33 + 0.53 = 10.26 \text{ t/m}$$

$$\Sigma R_T = \Sigma R_{OT} - \frac{X}{h} = 4.73 - \frac{9.83}{5.00} = 2.76 \text{ t/m}$$

$$\Sigma R_B = \Sigma R_{OB} + \frac{X}{h} = 10.26 + \frac{9.83}{5.00} = 12.23 = V_B \text{ t/m}$$

Moments:

$$M_{0,\max} = \frac{p_\gamma h^2}{9\sqrt{3}} = \frac{3.80 \cdot 5.00^2}{9 \cdot 1.73} = 6.10 \text{ t m/m}$$

$$x_0 = \frac{h}{\sqrt{3}} = \frac{5.00}{1.73} \cong 2.9 \text{ m}$$

$$M_{0,\max} = \frac{1}{8} 0.50 \cdot 0.422 \cdot 5.00^2 = 0.66 \text{ t m/m}, \ x_0 = 2.50 \text{ m}$$

$$\Sigma M_{0,m} = 2.70 + 6.10 + 0.66 = 9.46 \text{ t m/m}$$

$$z_{0m} = \frac{2.70 \cdot 1.81 + 6.10 \cdot 2.10 + 0.66 \cdot 2.50}{9.46} \cong 2.04 \text{ m}$$

$$M_{m,\max} = \Sigma M_{0m} - X \frac{z_{0m}}{h} = 9.46 - 9.83 \cdot \frac{2.04}{5.00} = 5.45 \text{ t m/m}$$

Footing:

$$\Sigma M_c = (X + \Sigma W_1\, 0.65 + W_2\, 0.0 - 2.73 \cdot 0.15) = -3.27 \text{ t m/m}$$

$$\Sigma W = 10.73 + 3.60 + 2.30 \cdot 0.40 \cdot 2.30 + 2.73 = 19.18 \text{ t/m}$$

$$e_0 = \frac{\Sigma M_c}{\Sigma W} = \frac{3.27}{19.18} = 0.17 \text{ m}$$

$$A = 1.0 \cdot 2 \cdot 1.15 = 2.30 \text{ m}^2$$

$$I_0 = \frac{2.30 \cdot 2.30^2}{12} = \sim 1.0 \text{ m}^4$$

Soil pressure:

$$p_{1,2} = -\frac{\Sigma w}{A} \pm \frac{\Sigma M_c}{I_0} a = -\frac{19.18}{2.30} \pm \frac{3.27}{1.00} \cdot 1.15$$

$$= -8.34 \pm 3.76 = -12.10 \text{ t/m}^2$$
$$-4.58 \text{ t/m}^2$$

$$W_{p1} = \frac{12.10 + 4.58}{2} 1.15 = 9.59 \text{ t/m}$$

$$W_{p2} = \frac{8.34 + 4.58}{2} \cdot 1.15 = 7.43 \text{ t/m}$$

$$c_1 = 0.61 \text{ m}, \ c_2 = 1.15 - 0.63 = 0.52 \text{ m}$$

Retaining Walls

$$M_c'' = W_{p1}c_1 = 9.59 \cdot 0.61 = 5.85 \text{ t m/m}$$

$$M_c' = -W_{p2}c_2 + (0.40 \cdot 1.15 \cdot 2.300 + 10.73)0.625 + 2.73 \cdot 0.15$$

$$= -7.43 \cdot 0.52 + 7.37 + 0.41 = 3.96 \text{ t/m}$$

$$\Sigma M_c \sim 0.0$$

Reinforcing and stresses:

Wall:

$$X = -9.83 \text{ tm}. \quad \Sigma P = 5.30 \text{ t}. \quad \sigma_c \cong 100 \text{ kg/cm}^2$$

$$x \cong 11.4 \text{ cm}, \quad z_0 \cong 19 \text{ cm}$$

$$T = -C = \frac{X}{z_0} = \frac{9.83}{0.19} = 51.74 \text{ t/m}$$

$$\sigma_c = -\left(\frac{2 \cdot 51.74}{11.4} + \frac{5.30}{30}\right)\frac{1000}{100} = -93 \text{ kg/cm}^2$$

$$A_s = \frac{9.83 + 0.07 \cdot 5.30}{0.19 \cdot 1.400} - \frac{5.30}{1.400} = 34.57 \text{ cm}^2$$

$$M_{m,\max} = 5.45 \text{ tm}, \quad N \cong 4.00 \text{ t m/m}$$

$$T = -C = \frac{5.45}{0.19} = 28.70 \text{ t/m}$$

$$A_s = \frac{28.7 - 4.00}{1.400} = 17.64 \text{ cm}^2$$

$$\sigma_c = -\left(\frac{2 \cdot 28.7}{11.4} + \frac{4.00}{30}\right)\frac{1000}{100} = 52 \text{ kg/cm}^2$$

$$v'' = \frac{V_B''}{bz_0} = \frac{12.23 \cdot 1000}{100 \cdot 19} \cong 6.4 \text{ kg/cm}^2$$

Footing:

$$M_c'' = 5.85 \text{ tm/m} \qquad k_3 = \frac{15 \cdot 50}{15 \cdot 50 + 1400} = 0.35$$

$$x = 7.7 \text{ cm}, \quad z_0 = 19.4 \text{ cm}$$

$$T_B = -C = \frac{M_c''}{z_0} = \frac{5.85}{0.194} \cong 30.2 \text{ t/m}$$

$$A_s = \frac{T_B}{\sigma_s} = \frac{30.2}{1.400} = 21.54 \text{ cm}^2 - \text{Bottom}$$

$$T_T = \frac{M'_c}{z_0} = \frac{3.96}{0.194} = 20.40 \text{ t/m}$$

$$A_s = \frac{T_T}{\sigma_s} = \frac{20.40}{1.400} = 14.58 \text{ cm}^2 - \text{Top}$$

Basement Walls Partly Below Grade

Wall: 30 cm concrete block openings filled with concrete. Reinforcing in the block openings.

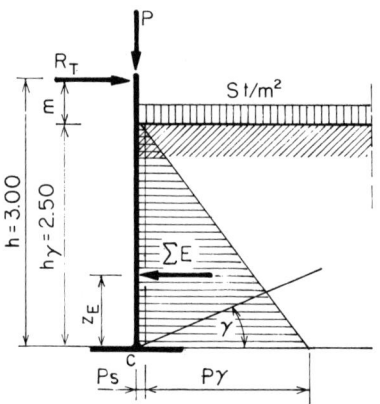

Figure 11-9

$$P = 4.87 \text{ t/m}$$
$$s = 0.250 \text{ t/m}^2$$
$$h = 3.00 \text{ m} \qquad \varphi = 22°$$
$$h_\gamma = 2.50 \text{ m} \qquad c = 0.455$$
$$\gamma = 2.00 \text{ t/m}^3$$
$$\gamma_c = 2.00 \text{ t/m}^3$$
$$E_s = sh_\gamma\, c = 0.250 \cdot 2.50 \cdot 0.455 = 0.284 \text{ t/m}$$
$$E_\gamma = \frac{1}{2}\gamma\, h_\gamma^2\, c = \frac{1}{2} \cdot 2.00 \cdot 2.50^2 \cdot 0.455 \cong 2.840 \text{ t/m}$$
$$p_s = \frac{E_s}{h_\gamma} = \frac{0.284}{2.50} = 0.114 \text{ t/m}^2$$

Retaining Walls

$$p_\gamma = \frac{2 \cdot E_\gamma}{h_\gamma} = \frac{2 \cdot 2.840}{2.50} = 2.272 \text{ t/m}^2$$

$$z_E = \frac{E_s \cdot h_\gamma/2 + E_\gamma \cdot h_\gamma/3}{E_s + E_\gamma} = 0.870 \text{ m}$$

$$R_{T0} = \frac{p_s \cdot h_\gamma^2}{2 \cdot h} = \frac{0.114 \cdot 2.50^2}{2 \cdot 3.00} = 0.119 \text{ t/m}, \quad R_{B0} = 0.166 \text{ t/m}$$

$$x_{sB} = \frac{R_{B0}}{p_s} = \frac{0.166}{0.114} = 1.46 \text{ m}$$

$$M_{0,\max} = \frac{R_{B0}^2}{2p_s} = \frac{0.166^2}{2 \cdot 0.114} = 0.12 \text{ t m/m}$$

$$R_{T0} = \frac{p_\gamma h_\gamma^2}{6h} = \frac{2.240 \cdot 2.50^2}{6 \cdot 3.00} = 0.78 \text{ t/m}, \quad R_{B0} = 2.03 \text{ t/m}$$

$$x_{0T} = m + h_\gamma \sqrt{\frac{h_\gamma}{3h}} = 0.50 + 2.50 \sqrt{\frac{2.50}{3 \cdot 3.00}} = \begin{matrix} 1.815 \text{ m} \\ (1.18) \end{matrix}$$

$$M_{0,\max} = R_{T0}\left(x_0 - \frac{h_\gamma}{3}\sqrt{\frac{h_\gamma}{3 \cdot h}}\right) = 0.780\left(1.815 - \frac{2.50}{3}\sqrt{\frac{2.50}{3 \cdot 3.00}}\right)$$

$$= 1.07 \text{ t m/m}$$

$$\Sigma M_{0c} = -(0.12 + 1.07) = -1.19 \text{ t m/m}$$

$$z_{0m} = \frac{0.12 \cdot 1.46 + 1.07 \cdot 1.18}{1.19} = 1.21 \text{ m}$$

Loading factors (Eq. 1-18):

$$\overline{L}_s'' = \frac{p_s h_\gamma^2}{h^2}\left(h - \frac{h_\gamma}{2}\right)^2 = \frac{0.114 \cdot 2.5^2}{3.00^2}\left(3.00 - \frac{2.50}{2}\right)^2 = 0.24$$

$$\overline{L}_\gamma'' = \frac{p_\gamma h_\gamma^2}{60\,h^2}(20h^2 - 15\,h\,h_\gamma + 3h_\gamma^2)$$

$$= \frac{2.240 \cdot 2.50^2}{60 \cdot 3.00^2}(20 \cdot 3.00^2 - 15 \cdot 3.00 \cdot 2.50 + 3 \cdot 2.50^2) = 2.23$$

$$\Sigma\overline{L}'' = 0.243 + 2.230 = 2.473 \text{ tm/m}$$

Moments:

$$\Sigma\overline{L}'' + 2X = 0 \qquad X = -\frac{2.473}{2} = -1.24 \text{ tm/m}$$

$$\Sigma M_{m,\max} = M_{0,\max} - \frac{X}{h} z_0 = 1.19 - \frac{1.24}{3.00} 1.77 = 0.46 \text{ t m/m}$$

$$R_T = \Sigma R_{T0} - \frac{X}{h} = 0.899 - \frac{1.24}{3.00} \cong 0.49 \text{ t/m}, \; R_B \cong 2.64 \text{ t/m}$$

Footing:

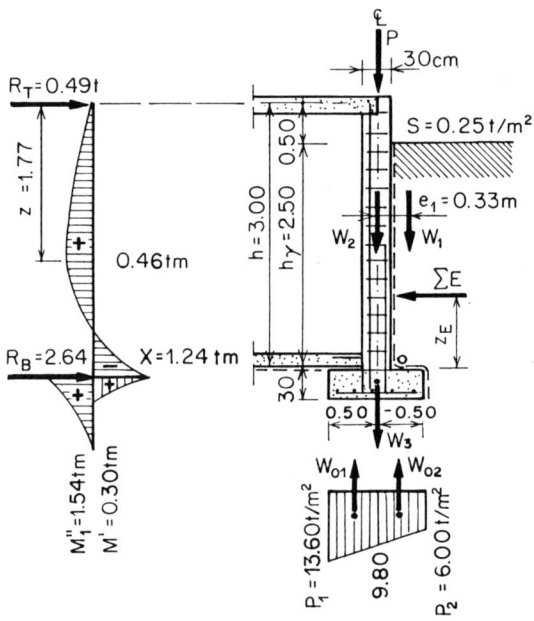

Figure 11-10

$$a = 2 \cdot 0.50 = 1.00 \text{ m}$$

$$d_0 = 30 \text{ cm}, \; d_0 - d' = 30 - 8 = 22 \text{ cm}$$

$$W_1 = (2.50 \cdot 1.60 + 0.250)(0.50 \cdot 0.15) = 1.49 \text{ t/m}$$

$$W_2 = 0.30 \cdot 2.50 \cdot 2.100 = 1.90 \text{ t/m}$$

$$E_v = 0.287 \cdot 2.84 = 0.82 \text{ t/m}$$

$$W_3 = 1.00 \cdot 0.30 \cdot 2.30 = 0.70 \text{ t/m}$$

$$\Sigma W = 4.87 + 1.49 + 1.90 + 0.82 + 0.70 = 9.78 \text{ t/m}$$

$$\Sigma M_c = -X + W_1 e_1 + E_v e_v$$

$$= -1.24 + 1.49 \cdot 0.33 + 0.82 \cdot 0.15 = -0.63 \text{ t m/m}$$

Retaining Walls

$$e_0 = \frac{\Sigma M_c}{\Sigma W} = \frac{0.63}{9.78} = \sim 0.06 \text{ m}$$

$$A = 1.00 \text{ m}^2$$

$$I_0 = \frac{1.00 \cdot 1.00^2}{12} = 0.083 \text{ m}^4$$

Soil pressure:

$$p_{1,2} = -\frac{\Sigma W}{A} \mp \frac{\Sigma M_c}{I_0} \cdot \frac{a}{2}$$

$$= -\frac{9.78}{1.00} \mp \frac{0.630}{0.083} \cdot 0.50 = \begin{array}{l} -13.60 \text{ t/m}^2 \\ -6.00 \text{ t/m}^2 \end{array}$$

$$W_{p1} = \frac{13.60 + 9.80}{2} 0.50 = 5.90 \text{ t/m}$$

$$c_1 = 0.26 \text{ m}$$

$$W_{p2} = \frac{9.80 + 6.00}{2} 0.50 = 4.00 \text{ t/m}$$

$$c_2 = 0.23 \text{ m}$$

$$M''_{p1} = 5.90 \cdot 0.26 = 1.53 \text{ t m/m}$$

$$M'_{p2} = -4.00 \cdot 0.23 + (0.484 + 0.123) = -0.30 \text{ t m/m}$$

$$\Sigma M = X + \Sigma M_p = -1.235 + (1.53 - 0.300) = \sim 0.0$$

Reinforcing and stresses:
Wall:

$$\sigma_c = 25 \text{ kg/cm}^2 \qquad \sigma_s = 1200 \text{ kg/cm}^2$$

$$k_x = \frac{15 \cdot 25}{15 \cdot 25 + 1200} = 0.24, \, d = 22 \text{ cm}$$

$$x = k_x \, d = 0.24 \cdot 22 \cong 5.2 \text{ cm}$$

$$z_0 = d - \frac{1}{3}x = 22 - 1.7 \sim 20 \text{ cm}$$

$$T = -C = \frac{X}{z_0} = \frac{1.235}{0.20} \cong 6.2 \text{ t/m}$$

$$A_s = \frac{6.2}{1.200} = 5.2 \text{ cm}^2/\text{m}$$

$$\sigma'_c = \frac{2C}{bx} = -\frac{2 \cdot 6.2}{100 \cdot 5.2} \cdot 1000 = -24 \text{ kg/cm}^2$$

Checking homogeneous material:

$$A = 30 \text{ cm}^2, \quad I_0 = \frac{30 \cdot 30^2}{12} = 2250 \text{ cm}^4$$

$$\sigma'_c = -\frac{\Sigma P}{A} \pm \frac{X}{I_0}\frac{a}{2} = \left(-\frac{4.87 + 1.90}{0.30} \pm \frac{1.24}{0.0025} 0.15\right)$$

$$= -22.57 \pm 80.87 = -103.4 \text{ t/m}^2 = -10.34 \text{ kg/cm}^2$$

$$+ 58.3 \text{ t/m}^2 = 5.83 \text{ kg/cm}^2$$

Use reinforcing

Footing:

$$M_{max} = M''_{p1} = 1.54 \text{ tm/m}$$

$$T = \frac{M''_{p1}}{z_0} = \frac{1.54}{0.20} = 7.70 \text{ t/m}$$

$$A_s = \frac{7.70}{1.200} = 6.4 \text{ cm}^2/\text{m} - \text{Bottom}$$

Appendix

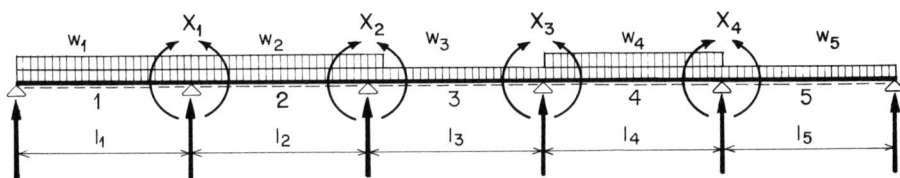

Figure A-1

TABLE 1.* Moment coefficients for continuous beam over three to six supports subjected to uniform various loading (w) at spans. $L_{min} = 0.80 L_{max}$, $c = 0.25$, $I =$ constant. Maximum moments for supports (X_n) and span moments (M_n).

Number of spans	Moment (max)	Spans participating				
		$w_1 l_1^2$	$w_2 l_2^2$	$w_3 l_3^2$	$w_4 l_4^2$	$w_5 l_5^2$
2	x_1	−0.0625	−0.0625			
	M_1	+0.0950	−0.0250			
	M_2	−0.0250	+0.0950			
3	x_1	−0.0625	−0.0500	+0.0156		
	x_2	+0.0156	−0.0500	−0.0625		
	M_1	+0.0950	−0.0200	+0.0063		
	M_2	−0.0234	+0.0750	−0.0234		
	M_3	+0.0063	−0.0200	+0.0950		
4	x_1	−0.0625	−0.0500	+0.0125	−0.0039	
	x_2	+0.0156	−0.0500	−0.0500	+0.0156	
	x_3	−0.0039	+0.0125	−0.0500	−0.0625	
	M_1	+0.0950	−0.0200	+0.0050	−0.0016	
	M_2	−0.0234	+0.0750	−0.0188	+0.0059	
	M_3	+0.0059	−0.0188	+0.0750	−0.0234	
	M_4	−0.0016	+0.0050	−0.0200	+0.0950	
5	x_1	−0.0625	−0.0500	+0.0125	+0.0031	+0.0010
	x_2	+0.0156	−0.0500	−0.0500	+0.0125	−0.0039
	x_3	−0.0039	+0.0125	−0.0500	−0.0500	+0.0156
	x_4	+0.0010	−0.0031	+0.0125	−0.0500	−0.0625
	M_1	+0.0950	−0.0200	+0.0050	−0.0013	+0.0004
	M_2	−0.0234	+0.0750	−0.0188	+0.0047	−0.0015
	M_3	+0.0059	−0.0188	+0.0750	−0.0188	+0.0059
	M_4	−0.0015	+0.0047	−0.0188	+0.0750	−0.0234
	M_5	+0.0004	−0.0013	+0.0050	−0.0200	+0.0950

*Tables 1 and 2 are taken from August Komendant's *Contemporary Concrete Structures* (New York: McGraw-Hill Book Company, 1972).

TABLE 2.* Coefficients for moments $X_{1\xi}$ Λ^{l}_{ξ} and reactions $R_{0\xi}$ of outside and an intermediate span subjected to a moving load $P = 1.0$. $c = 0.25$, L_{min} $0.80L_{max}$, $I =$ constant.

$\xi = \frac{x}{l}$, $\xi' = \frac{x'}{l}$

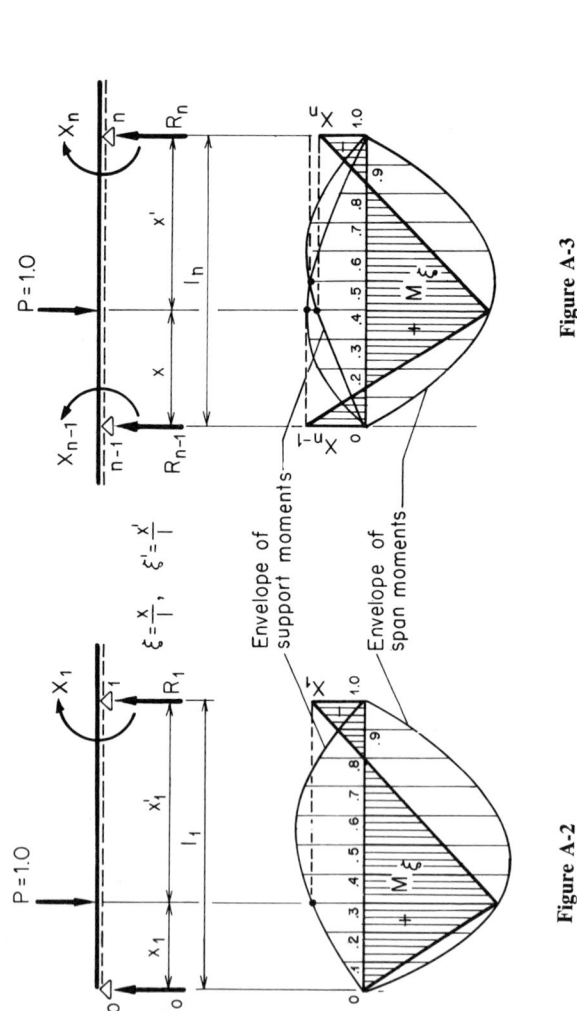

Figure A-2

Figure A-3

Appendix

End spans ($M = \alpha l$):*

ξ	X_1	$M_{1\xi}$	R_0	ξ	X_1	$M_{1\xi}$	R_0	ξ	X_1	$M_{1\xi}$	R_0
0.00	0.0000	0.0000	1.000	0.35	0.0763	0.2006	0.573	0.70	0.0892	0.1475	0.211
0.05	0.0125	0.0469	0.937	0.40	0.0840	0.2064	0.516	0.75	0.0820	0.1260	0.168
0.10	0.0248	0.0875	0.875	0.45	0.0897	0.2071	0.460	0.80	0.0720	0.1024	0.128
0.15	0.0366	0.1220	0.813	0.50	0.0937	0.2031	0.406	0.85	0.0590	0.0774	0.091
0.20	0.0480	0.1504	0.752	0.55	0.0959	0.1948	0.354	0.90	0.0427	0.0515	0.057
0.25	0.0586	0.1728	0.691	0.60	0.0960	0.1824	0.304	0.95	0.0231	0.0255	0.027
0.30	0.0682	0.1895	0.632	0.65	0.0938	0.1665	0.256	1.00	1.0000	0.0000	0.000

Intermediate spans ($M = \alpha l$):*

ξ	X_{n-1}	X_n	M_n	R_{n-1}	ξ	X_{n-1}	X_n	M_n	R_{n-1}
0.00	0.0000	0.0000	0.0000	1.000	0.55	0.0701	0.0784	0.1728	0.442
0.05	0.0214	0.0071	0.0268	0.964	0.60	0.0640	0.0800	0.1664	0.384
0.10	0.0390	0.0150	0.0534	0.924	0.65	0.0569	0.0796	0.1558	0.328
0.15	0.0531	0.0234	0.0788	0.880	0.70	0.0490	0.0770	0.1414	0.272
0.20	0.0640	0.0320	0.1024	0.832	0.75	0.0406	0.0719	0.1234	0.219
0.25	0.0719	0.0406	0.1234	0.781	0.80	0.0320	0.0640	0.1024	0.168
0.30	0.0770	0.0490	0.1414	0.728	0.85	0.0234	0.0531	0.0788	0.120
0.35	0.0796	0.0569	0.1558	0.673	0.90	0.0150	0.0390	0.0534	0.076
0.40	0.0800	0.0640	0.1664	0.616	0.95	0.0071	0.0214	0.0268	0.036
0.45	0.0784	0.0701	0.1728	0.558	1.00	0.0000	0.0000	0.0000	0.000
0.50	0.0750	0.0750	0.1750	0.500					

*If the location of load application differs from ξ, interpolation can be used.

TABLE 3. Support moments and slab moment coefficients

Slab type a:

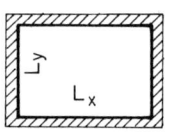

$$M_{x,\max} = c_{ax} L_x^2 \qquad w_x = k_{ox} w$$
$$M_{y,\max} = c_{ay} L_y^2 \qquad w_y = k_{oy} w$$

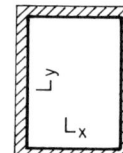

ε L_y/L_x	c_{ax}	c_{ay}	k_{ox}	k_{oy}	ε L_y/L_x	c_{ax}	c_{ay}	k_{ox}	k_{oy}
0.60	0.011	0.081	0.115	0.885	1.00	0.036	0.036	0.500	0.500
0.65	0.013	0.074	0.152	0.848	1.05	0.040	0.033	0.549	0.451
0.70	0.016	0.068	0.194	0.806	1.10	0.044	0.030	0.594	0.406
0.75	0.019	0.061	0.240	0.760	1.15	0.048	0.027	0.636	0.364
0.80	0.023	0.055	0.291	0.709	1.20	0.051	0.025	0.675	0.325
0.85	0.026	0.050	0.343	0.657	1.30	0.059	0.021	0.741	0.259
0.90	0.029	0.045	0.396	0.604	1.40	0.066	0.017	0.794	0.206
0.95	0.033	0.040	0.449	0.551	1.50	0.072	0.014	0.835	0.165

Slab type b:

$$M_{x,\max} = c_{bx} w L_x^2 \qquad w_x = k_{ox} w$$
$$M_{y,\max} = c_{by} w L_y^2 \qquad w_y = k_{oy} w$$

ε L_y/L_x	c_{bx}	c_{by}	k_{ox}	k_{oy}	ε L_y/L_x	c_{bx}	c_{by}	k_{ox}	k_{oy}
0.60	0.004	0.034	0.115	0.885	1.00	0.018	0.018	0.500	0.500
0.65	0.006	0.032	0.152	0.848	1.05	0.020	0.016	0.549	0.451
0.70	0.007	0.030	0.194	0.806	1.10	0.021	0.015	0.594	0.406
0.75	0.009	0.028	0.240	0.760	1.15	0.023	0.013	0.636	0.364
0.80	0.011	0.026	0.291	0.709	1.20	0.024	0.012	0.675	0.325
0.85	0.012	0.024	0.343	0.657	1.30	0.027	0.009	0.741	0.259
0.90	0.014	0.022	0.396	0.604	1.40	0.029	0.008	0.794	0.206
0.95	0.016	0.020	0.449	0.551	1.50	0.031	0.006	0.835	0.165

$$X_x = -\frac{1}{12} w_x L_x^2$$
$$X_y = -\frac{1}{12} w_y L_y^2$$

Slab type c:

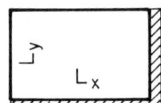

$$M_{x,\max} = c_{cx}\,w\,L_x^2 \qquad w_x = k_{ox}\,w$$
$$M_{y,\max} = c_{cy}\,w\,L_y^2 \qquad w_y = k_{oy}\,w$$
$$X_x = \frac{1}{8}w_x L_x^2 \qquad X_y = -\frac{1}{8}w_y L_y^2$$

ε L_y/L_x	c_{cx}	c_{cy}	k_{ox}	k_{oy}	ε L_y/L_x	c_{cx}	c_{cy}	k_{ox}	k_{oy}
0.60	0.007	0.053	0.115	0.885	1.00	0.027	0.027	0.500	0.500
0.65	0.009	0.050	0.152	0.848	1.05	0.030	0.024	0.549	0.451
0.70	0.011	0.046	0.194	0.806	1.10	0.032	0.022	0.594	0.406
0.75	0.014	0.043	0.240	0.760	1.15	0.035	0.020	0.636	0.364
0.80	0.016	0.039	0.291	0.709	1.20	0.037	0.018	0.675	0.325
0.85	0.019	0.036	0.343	0.657	1.30	0.041	0.014	0.741	0.259
0.90	0.021	0.033	0.396	0.604	1.40	0.045	0.012	0.794	0.206
0.95	0.024	0.030	0.449	0.551	1.50	0.049	0.010	0.835	0.165

Slab type 1:

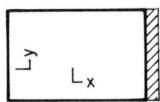

$$M_{xc,\max} = c_{1x}\,w\,L_x^2 \qquad w_x = k_{1x}\,w$$
$$M_{y,\max} = c_{1y}\,w\,L_y^2 \qquad k_y = k_{1y}\,w$$
$$X_x = -\frac{1}{8}w_x L_x^2$$

ε L_y/L_x	c_{1x}	c_{1y}	k_{1x}	k_{1y}	ε L_y/L_x	c_{1x}	c_{1y}	k_{1x}	k_{1y}
0.60	0.012	0.073	0.245	0.755	1.00	0.033	0.027	0.714	0.286
0.65	0.014	0.065	0.309	0.691	1.05	0.036	0.024	0.752	0.248
0.70	0.017	0.058	0.375	0.625	1.10	0.038	0.021	0.785	0.215
0.75	0.020	0.052	0.442	0.558	1.15	0.041	0.018	0.814	0.186
0.80	0.022	0.045	0.506	0.494	1.20	0.043	0.016	0.838	0.162
0.85	0.025	0.040	0.566	0.434	1.30	0.047	0.013	0.877	0.123
0.90	0.028	0.035	0.621	0.379	1.40	0.050	0.010	0.906	0.094
0.95	0.031	0.031	0.671	0.329	1.50	0.053	0.008	0.927	0.073
ε' L_x/L_y	c_{1y}	c_{1x}	k_{1y}	k_{1x}	ε' L_x/L_y	c_{1y}	c_{1x}	k_{1y}	k_{1x}

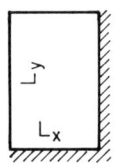

$$M_{x,\max} = c_{1x}\,w\,L_x^2 \qquad w_x = k_{1x}\,w$$
$$M_{y,\max} = c_{1y}\,w\,L_y^2 \qquad w_y = k_{1y}\,w$$
$$X_y = -\frac{1}{8}w_y L_y^2$$

Slab type 2:

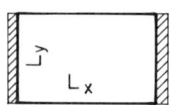
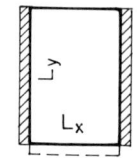

$$M_{x,\max} = c_{2x}\, w\, L_x^2 \qquad w_x = k_{2x}\, w$$
$$M_{y,\max} = c_{2y}\, w\, L_y^2 \qquad w_y = k_{2y}\, w$$
$$X_x = -\frac{1}{12}\, w_x\, L_x^2$$

ε L_y/L_x	c_{2x}	c_{2y}	k_{2x}	k_{2y}	ε L_y/L_x	c_{2x}	c_{2y}	k_{2x}	k_{2y}
0.60	0.011	0.062	0.393	0.607	1.00	0.027	0.018	0.833	0.167
0.65	0.014	0.054	0.472	0.528	1.05	0.028	0.015	0.859	0.141
0.70	0.016	0.046	0.546	0.454	1.10	0.029	0.013	0.880	0.120
0.75	0.018	0.040	0.613	0.387	1.15	0.030	0.011	0.897	0.103
0.80	0.020	0.034	0.672	0.328	1.20	0.031	0.010	0.912	0.088
0.85	0.022	0.029	0.723	0.277	1.30	0.033	0.007	0.935	0.065
9.90	0.024	0.025	0.766	0.234	1.40	0.034	0.006	0.951	0.049
0.95	0.025	0.021	0.803	0.197	1.50	0.035	0.004	0.962	0.038
ε' L_x/L_y	c_{2y}	c_{2x}	k_{2y}	k_{2x}	ε' L_x/L_y	c_{2y}	c_{2x}	k_{2y}	k_{2x}

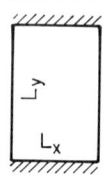
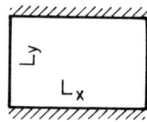

$$M_{x,\max} = c_{2x}\, w\, L_x^2 \qquad w_x = k_{2x}\, w$$
$$M_{y,\max} = c_{2y}\, w\, L_y^2 \qquad w_y = k_{2y}\, w$$
$$X_y = -\frac{1}{12}\, w_y\, L_y^2$$

Slab type 3:

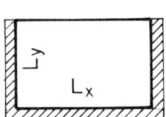
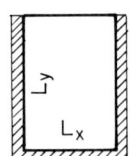

$$M_{x,\max} = k_{3x}\, w\, L_x^2 \qquad w_x = k_{3x}\, w$$
$$M_{y,\max} = k_{3y}\, w\, L_y^2 \qquad w_y = k_{3y}\, w$$
$$X_x = -\frac{1}{12}\, w_x\, L_x^2 \qquad X_y = -\frac{1}{8}\, w_y\, L_y^2$$

ε L_y/L_x	c_{3x}	c_{3y}	k_{3x}	k_{3y}	ε L_y/L_x	c_{3x}	c_{3y}	k_{3x}	k_{3y}
0.60	0.007	0.048	0.206	0.794	1.00	0.023	0.020	0.667	0.333
0.65	0.009	0.044	0.263	0.737	1.05	0.024	0.017	0.709	0.291
0.70	0.011	0.040	0.324	0.676	1.10	0.026	0.015	0.745	0.255
0.75	0.013	0.036	0.388	0.612	1.15	0.027	0.013	0.778	0.222
0.80	0.015	0.032	0.450	0.550	1.20	0.028	0.012	0.806	0.194
0.85	0.017	0.029	0.511	0.489	1.30	0.031	0.009	0.851	0.149
0.90	0.019	0.025	0.567	0.433	1.40	0.032	0.007	0.885	0.115
0.95	0.021	0.022	0.620	0.380	1.50	0.034	0.006	0.910	0.090
ε' L_x/L_y	c_{3y}	c_{3x}	k_{3y}	k_{3x}	ε' L_x/L_y	c_{3y}	c_{3x}	k_{3y}	k_{3x}

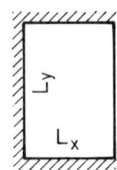
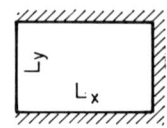

$$M_{x,\max} = c_{3x}\, w\, L_x^2 \qquad w_x = k_{3x}\, w$$
$$M_{y,\max} = c_{3y}\, w\, L_y^2 \qquad w_y = k_{3y}\, w$$
$$X_x = -\frac{1}{8}\, w_x\, L_x^2 \qquad X_y = -\frac{1}{12}\, w_y\, L_y^2$$

TABLE 4. SOIL PHYSICAL CHARACTERISTICS

Soil type	t/m^3	μ	c	p t/m^2	k^* kg/cm^{-3}
Fine loose sand Moderately dry Saturated	1.6–1.9 1.9–2.1	31	0.320	16–20	36–58
Dense sand Moderately dry Saturated	1.9–2.0 2.1–2.2	32	0.307	25–35	50–100
Dense Gravel-sand mixture Moderately dry	2.0–2.2	33	0.295	35–40	70–125
Poorly graded Gravel-sand mixture Moderately dry	1.9–2.0	30	0.333	35–60	54–90
Well graded Gravel-sand mixture Moderately dry	2.2–2.3	33	0.295	55–70	90–125
Clay and sand mixture	1.8–2.2	22 26	0.455 0.390	15–25	30–58
Soft clay-silt mixture	1.5–2.0	12 16	0.656 0.563	5–20	10–27
Soft rock Sound rock		– –	– –	80–160 200–400	– –

$$*k = \frac{p'}{\delta} = \frac{\text{pressure}}{\text{deflection}}$$

TABLE 5. ELEMENTARY TRANSCENDENTAL FUNCTIONS

x	x, deg	e^x	e^{-x}	$\sin x$	$\cos x$	$\tan x$
0.00	0.00	1.00000	1.00000	0.00000	1.00000	0.00000
0.01	0.58	1.01010	0.99005	0.01000	0.99995	0.01000
0.02	1.15	1.02020	0.98020	0.02000	0.99980	0.02000
0.03	1.72	1.03050	0.97045	0.03000	0.99955	0.03001
0.04	2.28	1.04080	0.96079	0.03999	0.99920	0.04002
0.05	2.87	1.05130	0.95123	0.04998	0.99875	0.05004
0.06	3.43	1.06180	0.94176	0.05996	0.99820	0.06007
0.07	4.00	1.07250	0.93239	0.06994	0.99755	0.07011
0.08	4.58	1.08330	0.92312	0.07991	0.99680	0.08017
0.09	5.17	1.09420	0.91393	0.08988	0.99595	0.09024
0.1	5.73	1.10517	0.90484	0.09983	0.99500	0.10033
0.2	11.46	1.22140	0.81873	0.19867	0.98007	0.20271
0.3	17.19	1.34986	0.74082	0.29552	0.95534	0.30934
0.4	22.92	1.49182	0.67032	0.38942	0.92106	0.42279
0.5	28.65	1.64872	0.60653	0.47943	0.87758	0.54630
0.6	34.38	1.82212	0.54881	0.56464	0.82534	0.68414
0.7	40.11	2.01375	0.49659	0.64422	0.76484	0.84229
0.8	45.84	2.22554	0.44933	0.71736	0.69671	1.02964
0.9	51.57	2.45960	0.40657	0.78333	0.62161	1.26016
1.0	57.30	2.71828	0.36788	0.84147	0.54030	1.55741
1.1	63.03	3.00417	0.33287	0.89121	0.45360	1.96476
1.2	68.75	3.32012	0.30119	0.93204	0.36236	2.57215
1.3	74.48	3.66930	0.27253	0.96356	0.26750	3.60210
1.4	80.21	4.05520	0.24660	0.98545	0.16997	5.79789
1.5	85.94	4.48169	0.22313	0.99749	+0.07074	+14.10142
1.6	91.67	4.95303	0.20190	0.99957	−0.02920	−34.23254
1.7	97.40	5.47395	0.18268	0.99166	−0.12884	−7.69660
1.8	103.13	6.04965	0.16530	0.97385	−0.22720	−4.28626
1.9	108.86	6.68589	0.14957	0.94630	−0.32329	−2.92710
2.0	114.59	7.38906	0.13534	0.90930	−0.41615	−2.18504
2.1	120.32	8.16617	0.12246	0.86321	−0.50485	−1.70985
2.2	126.05	9.02501	0.11080	0.80850	−0.58850	−1.37382
2.3	131.78	9.97418	0.10026	0.74571	−0.66628	−1.11921
2.4	137.51	11.02318	0.09072	0.67546	−0.73739	−0.91601
2.5	143.24	12.18249	0.08208	0.59847	−0.80114	−0.74702
2.6	148.97	13.46374	0.07427	0.51550	−0.85689	−0.60160
2.7	154.70	14.87973	0.06721	0.42738	−0.90407	−0.47273

Appendix

x	x, deg	e^x	e^{-x}	$\sin x$	$\cos x$	$\tan x$
2.8	160.43	16.44465	0.06081	0.33499	−0.94222	−0.35553
2.9	166.16	18.17415	0.05502	0.23925	−0.97096	−0.24641
3.0	171.89	20.08554	0.04979	0.14112	−0.98999	−0.14255
3.1	177.62	22.19795	0.04505	+0.04158	−0.99914	−0.04162
3.2	183.35	24.53253	0.04076	−0.05837	−0.99829	+0.05847
3.3	189.08	27.11264	0.03688	−0.15775	−0.98748	0.15975
3.4	194.81	29.96410	0.03337	−0.25554	−0.96680	0.26442
3.5	200.54	33.11545	0.03020	−0.35078	−0.93646	0.37470
3.6	206.26	36.59823	0.02732	−0.44252	−0.89676	0.49347
3.7	211.99	40.44730	0.02472	−0.52984	−0.84810	0.62473
3.8	217.72	44.70118	0.02237	−0.61186	−0.79097	0.77356
3.9	223.45	49.40245	0.02024	−0.68777	−0.72593	0.94742
4.0	229.18	54.59815	0.01832	−0.75680	−0.65364	1.15782
4.1	234.91	60.34029	0.01657	−0.81828	−0.57482	1.42353
4.2	240.64	66.68633	0.01500	−0.87158	−0.49026	1.77778
4.3	246.37	73.69979	0.01357	−0.91617	−0.40080	2.28585
4.4	252.10	81.45087	0.01228	−0.95160	−0.30733	3.09632
4.5	257.83	90.01713	0.01111	−0.97753	−0.21080	4.63733
4.6	263.56	99.48432	0.01005	−0.99369	−0.11215	8.86018
4.7	269.29	109.9472	0.00910	−0.99992	−0.01239	+80.71280
4.8	275.02	121.5104	0.00823	−0.99616	+0.08750	−11.38487
4.9	280.75	134.2898	0.00745	−0.98245	0.18651	−5.26749
5.0	286.48	148.4132	0.00674	−0.95892	0.28366	−3.38052
5.1	292.21	164.0219	0.00610	−0.92581	0.37798	−2.44939
5.2	297.94	181.2722	0.00552	−0.88345	0.46852	−1.88564
5.3	303.67	200.3368	0.00499	−0.83227	0.55437	−1.50128
5.4	309.40	221.4064	0.00452	−0.77276	0.63469	−1.21754
5.5	315.13	244.6919	0.00409	−0.70554	0.70867	−0.99558
5.6	320.86	270.4264	0.00370	−0.63127	0.77557	−0.81394
5.7	326.59	298.8674	0.00335	−0.55069	0.83471	−0.65973
5.8	332.32	330.2996	0.00303	−0.46460	0.88552	−0.52467
5.9	338.05	365.0375	0.00274	−0.37388	0.92748	−0.40311
6.0	343.77	403.4288	0.00248	−0.27942	0.96017	−0.29101
6.1	349.50	445.8578	0.00224	−0.18216	0.98327	−0.18526
6.2	355.23	492.7490	0.00203	−0.08309	0.99654	−0.08338
6.3	360.96	544.5719	0.00184	+0.01681	0.99986	+0.01682

TABLE 6. STANDARD U.S. REINFORCING BARS: A15, A432, A431

No.	Diameter		Area		Perimeter		Weight	
	in.	mm	in.^{2}	cm^2	in.	cm	lb/ft	kg/m
2	1/4	6.35	0.05	0.32	0.785	1.99	0.167	0.248
3	3/8	9.53	0.11	0.71	1.178	2.99	0.376	0.559
4	1/2	12.70	0.20	1.29	1.571	3.99	0.668	0.994
5	5/8	15.87	0.31	2.00	1.963	4.99	1.043	1.552
6	3/4	19.05	0.44	2.84	2.356	5.98	1.502	2.235
7	7/8	22.23	0.60	3.87	2.749	6.98	2.044	3.042
8	1	25.40	0.79	5.10	3.142	7.98	2.670	3.973
9	1+	28.23	1.00	6.45	4.000	10.16	3.400	5.060
10	1 1/8	28.57	1.27	8.19	4.500	11.43	4.303	6.404
11	1 1/4	31.75	1.56	10.06	5.000	12.70	5.313	7.907

Modulus of elasticity: $E = 30 \times 10^6$ lbs/in.$^2 = 2.1 \times 10^6$ kg/cm^2 $\sigma_y = 40$ k/in.$^2 = 2.81$ t/cm^2; 60 k/in.$^2 = 4.22$ t/cm^2; 75 k/in.$^2 = 5.27$ t/cm^2

TABLE 7. COMMON-TYPE WELDED WIRE FABRIC (TWO-WAY $L = T$)

Type	Spacing		Wires gage	Sectional area		Weight	
	in.	cm	W & M	in.^{2}	cm^2	lb/100 ft^2	kg/10 m^2
4×4–4/4	4	10	4	0.120	2.547	85	41.50
4×4–6/6	4	10	6	0.087	1.845	62	30.30
4×4–8/8	4	10	8	0.062	1.316	44	21.48
4×4–10/10	4	10	10	0.043	0.912	31	15.13
4×4–12/12	4	10	12	0.026	0.552	19	9.28
6×6–0/0	6	15	0	0.148	3.141	107	54.05
6×6–2/2	6	15	2	0.108	2.292	78	38.08
6×6–4/4	6	15	4	0.080	1.598	58	28.32
6×6–6/6	6	15	6	0.058	1.231	42	20.51
6×6–8/8	6	15	8	0.041	0.874	30	14.65

Width of standard fabrics = 5.00 ft = 1.52 m
$\sigma_y = 75$ k/in.$^2 = 5.27$ t/cm^2
$\sigma_s = 24$ to 28 k/in.$^2 = 1.700$ to 2.00 t/cm^2

TABLE 8. MINIMUM BREAKING STRENGTH OF SEVEN-WIRE STRANDS

Diameter		Nominal area		Min. breaking strength		Approximate weight of strands	
in.	mm	in.^{2}	cm^2	k	t	lb/ft	kg/m
3/8	9.52	0.085	0.551	23.0	10.470	0.292	0.434
1/2	12.70	0.153	0.987	41.3	18.733	0.525	0.780
0.6	15.24	0.217	1.378	58.6	26.182	0.740	1.101

Modulus of elasticity: $E = 28 \times 10^6$ lb/in.$^2 = 1.95 \times 10^6$ kg/cm^2 $\sigma_{uL} = 270$ kips/in.$^2 = 19.0$ t/cm^2

Appendix

TABLE 9. ENGINEERING CONVERSION FACTORS

Multiply	By	To obtain
Inches	2.54001	Centimeters
Inches	2.54001×10^{-2}	Meters
Inches	25.4001	Millimeters
Feet	30.4801	Centimeters
Feet	0.304801	Meters
Feet	304.801	Millimeters
Yards	0.914402	Meters
Square inches	645.163	Square millimeters
Square inches	6.45163	Square centimeters
Square feet	0.0929034	Square meters
Square feet	9.29034×10^{-6}	Hectares
Square feet	929.034	Square centimeters
Square yards	0.83613	Square meters
Cubic inches	16.38716	Cubic centimeters
Cubic feet	2.8317×10^{4}	Cubic centimeters
Cubic feet	2.8317×10^{-2}	Cubic meters
Cubic yards	0.764559	Cubic meters
Gallons, U.S.	0.13368	Cubic feet
Gallons, U.S.	231.	Cubic inches
Gallons, U.S.	3.78543	Liters
Gallons, British Imperial	1.20091	Gallons, U.S.
Pounds, avoirdupois	453.592	Grams, metric
Pounds, avoirdupois	0.453592	Kilograms
Pounds, avoirdupois	4.53592×10^{-4}	Tons, metric
Kips	453.592	Kilograms
Tons, short	907.185	Kilograms
Tons, long	1,016.05	Kilograms
Cubic feet, water	28.317	Kilograms
Pounds per foot	1.48816	Kilograms per meter
Pounds per square foot (psf)	4.88241	Kilograms per square meter
Pounds per square inch (psi)	7.031×10^{-2}	Kilograms per square centimeter
Pounds per square inch (psi)	7.031×10^{-4}	Kilograms per square millimeter
Pounds per cubic foot	16.0184	Kilograms per cubic meter
Foot-pound (ft-lb)	0.1383	Kilogram meters

TABLE 10. SI-SYSTEM CONVERSION

$k = 1,000 \qquad M = k^2 = 1,000,000$

1 kg = 10 N	1 N = 0.1 kp
= 0.01 kN	1 kN = 100 kg
1 t = 0.01 MN	1 MN = 100 t = 100 Mp
1 kgm = 0.01 kNm	1 kNm = 100 kgm = 0.1 tm
1 tcm = 0.01 MNcm	1 MNcm = 100 tcm = 100 Mpcm
1 tm = 0.01 MNm	1 MNm = 100 tm
1 kg/cm² = 0.10 MN/m²	1 MN/m² = 10 kg/cm²
1 t/cm² = 100 MN/m²	1 MN/m² = 0.01 t/cm²
1 kg/m² = 0.01 kN/m²	1 kN/m² = 100 kg/m²
1 kg/m³ = 0.01 kN/m³	1 kN/m³ = 100 kg/m³

Index

Acceleration, gravity, 212
Allowable stresses:
   concrete, 118
   reinforcing steel, 118, 250
   strands, 118, 250
   welded wire fabric, 250
Anchorage, 154
Anchorage loss, 147
Apparent modulus of elasticity, 117
Application of prestressing, 135
Arch action:
   curvature, 162
   elastic-plastic deformations, 165, 166
   fixed, 172–175
   principles of analysis, 160–163, 168
   shapes, 166, 171
   span-rise ratio, 162
   temperature change, 165, 166
   three-hinged, 164, 165
   two-hinged, 167–171
Arrangement of tendons:
   beams, slabs, 139–145, 152
   folded plates, 184–187
   trusses, 83–85
Assumptions and rules, 3, 4

Base shear, 211
Beams:
   simple, statically determined:
      deflection, 10–13
      end rotations, 18–21
      moments, reactions, shear, 9–13
      prestressing, 138–145
   statically indetermined:
      carryovers, 25, 26
      continuity requirements, 14, 24, 139, 141–143
      critical live load locations, 34, 35
      loading factor method, 19–23
      moment coefficients, 241–243
      moments in short spans, 38, 39
      prestressing, 132–135
      principal system, 14
      reaction coefficients, 16, 243

relationship, carryover, and zero-point distance, 23, 25
span-by-span loading, 24, 29–33, 47
span-depth ratio, 17, 118, 119
superposition, 24, 52
support moments, 26, 27
three moments equation, 24, 26
zero-point locations, 15, 26, 47
zero-point methods, 15
Bearing neoprene, 71, 72, 218
Bond stress, 4, 117, 118
Boundary conditions, 176, 182, 188
Buckling, 162

Carrying capacity, 128, 129
Carryovers, 25, 26, 48, 49
Center of gravity, 5–7
Centerline of arches, 162
Centerline of curvilinear shells (barrels), 187
Coefficient of thermal expansion, 68
Column moments, 49, 50, 53, 71, 72, 216–219
Column strips, 113
Continuity requirements, 14, 24, 139, 141–143
Coulomb's theory, 227
Cross-sectional coefficients, 5–7
Crown deflection, 165
Cycloid curvature, 187, 191, 192

Deflections:
 beams, elastic-plastic, 10–13, 98, 105, 158, 159
 beams or slabs on semi-elastic soil, 223–225
 lateral of frames, 65–67
 vierendeels, 86–95
Deformations, 8, 15, 42–46
Depth-span ratio. *See* span-depth ratio
Deterioration factor, 213
Distribution factors, 47, 48
Domes:
 polygonal:
  edge members, 201
  membrane forces, 198–205
  meridian curvatures, 201, 204
  types, 199

Domes:
 rotation symmetrical:
  deformations, 181, 182
  edge members, 181, 182
  Guldin's rule, 177
  membrane forces, 177
  meridian curvatures, 176
  temperature change, 181
Ductility, 213, 214

Effective width of slab in T-beams, 5
Elastic limits:
 concrete, 116
 reinforcing steel, 118
 seven-wire strands, 118
Elastically controlled joints, 70, 71, 214, 218, 219
End rotations, 9–13, 18–21
Energy method, 211–214
Engineering conversion factors, 251

Factor of safety, 117, 145, 146
Failure of structures, 129
Flat slabs, 112–115
Floors:
 two-skin, 114, 115
 waffle type, 114, 115
Folded plates, 184–187
Forces:
 earthquake, 210–214
 external, 3
 internal, 3
 prestressing, 132–135
 wind, 214, 215
Frames:
 carryovers, 25, 26, 48
 column moments, 49, 50, 53
 distribution factors, 47, 48
 horizontal loading, 55–65
 loading factors, 19–23
 span-by-span loading, 47
 span moments, 49, 52
 stiffnesses, 47
 superposition, 24, 52
 support moments, 27, 49
 temperature change and shrinkage, 68–71
 types, 41
 zero-point method, 42–46, 55–65

Index

Frequencies:
   circular, 215
   natural, 215
Frictional coefficients, 147
Frictional loss, 147, 148, 152
Funicular centerline, 162

Gerber-type girders, 88, 159
Grid systems, 103, 104
Grouting, 135
Guldin's rule, 177, 199

Hooke's law, 2, 149

Inflection points, 15
Initial losses, 147
   anchorage, 147
   frictional, 147, 148

K-values of soil, 222, 247
Kinetic energy, 211

Lateral deflection of frames, 65, 67
Losses in prestressing, 147–149
Lundgren method, 193

Magnification factor, 215
Marcus' method, 97
Marginal members. *See* Shells
Membrane forces. *See* Shells
Meridian curvatures, 177
Middle-strip or two-way slabs, 104, 112, 113
Modulus of elasticity:
   concrete, 117
   steel, 117
   strands, 250
Moor's theory, 9, 19

Noeprene bearings and interfaces, 70, 214, 218, 219
Neutral plane, 2, 216
Noise propagation, 214

Paraboloids, 205–209
Plastic deformations, 149
Poisson's ratio, 181
Polygonal domes, 198–205

Pressure:
   hydraulic, 227
   soil, 222, 227, 236
   wind, 214
Prestressing:
   force, 135
   losses, 155
   posttensioning, 135
   prestresses, 135
   prestressed continuous beams, 139–145
   pretensioning, 135
   suspension action, 132–137
   tendons' arrangement, 132–146, 152
Principal system, 14, 15

Radius of gyration, 134
Relationships:
   external loads and internal stresses, 4
   zero-point distances and carryovers, 15, 25, 26
Relative stiffnessess, 47
Reserve energy, 211, 212, 215
Resonance, 215
Ridge loads, 184, 185, 199–201
Rise of arches, 162
ROS, 116

Secondary stresses, 83, 176, 188, 192, 193
Shape of arches. *See* Arch action
Shear, 4, 9–12, 55–65, 188–192
Shear strength, 117
Shells, curvilinear:
   curvatures, 187
   circular, 190, 191
   cycloidal, 191, 192
   elliptical, 189, 190
   marginal members, 192
   membrane forces, 189–192
Shrinkage, 68–70, 148, 149
Slenderness, 105, 115, 118, 119
Soil characteristics, 247
Span-by-span loading, 24, 27, 29–33
Span-depth ratio, 17, 105, 115
Span-rise ratio of arches. *See* Arch action
Spectral values:
   acceleration, 212
   velocity, 212
Steel, strands, 250

Stiffness factors, 47, 212
Stresses, 2, 130, 131
Superposition, 2, 24, 49
Support moments, 27, 49

Three-hinged arches. *See* Arch action
Trusses, 83–86
Two-hinged arches. *See* Arch action
Two-way slabs, 97–103

Ultimate load, 129
Ultimate strength:
   concrete, 117, 118, 130, 131
   strands, 250

Velocity, 212, 214, 215
Vibration, 17
Vierendeels, 86–95

Water pressure, 227
Wind, 214, 186

Yield stress:
   reinforcing bars, 118, 250
   welded wire fabric, 250

Zero-point distances, 15, 47, 55, 223

RAYMOND H. FOGLER LIBRARY
DATE DUE

BOOKS ARE SUBJECT TO
RECALL AFTER TWO WEEKS